Atomic Coherence and its Potential Applications

Editor: Jin-Yue Gao

Co-Editor: Min Xiao
Yifu Zhu

CONTENTS

FOREWORD

In quantum mechanics it is known that if a transition takes place from an initial state to a final state via several different paths then one should add coherently all transition amplitudes before calculating the expression for transition probability. The coherent addition leads to quantum interferences and these are quite ubiquitous for quantum systems. Further any interactions in the final state would significantly affect the transitions rates. This was first extensively discussed by Fano and led to flurry of activity both theoretical and experimental. Such interferences are even more pronounced for higher order processes. Early experiments on multi photon processes and more specifically on the competition of one photon and three photon processes showed very remarkable role of interferences. The coherent path ways naturally lead to the production of atomic coherences. About two decades ago the scientific community realized the great potential of quantum interferences largely due to the work of Steve Harris and collaborators. By now we have seen many important applications of quantum interferences and atomic coherences and many of these applications have been thoroughly reviewed. However newer and newer applications continue to appear and one needs a good discussion at one place rather than going through the large volume of papers.

The current book by Gao and collaborators fills a gap and provides a comprehensive discussion of a variety of important new applications. These for example include- methods to reach the white cavity limit; production of nano scale resolution for microscopy and lithography applications; production of controllable photonic band gaps; creation of memory elements.

The material has been prepared by authors who themselves have extensive contributions to the field. The book is timely and should be quite useful for young researchers and those who like to get a feel for the great potential of quantum interferences and atomic coherences.

I enjoyed the book and am happy to have such a reference on my shelf.

Girish Agarwal, FRS

Noble Foundation Chair and Regents Professor
Oklahoma State University
USA

Dec, 2009

PREFACE

Laser-induced atomic coherence and interference, and related effects in multi-level atomic and molecular systems have been actively studied in 70's and 80's as important techniques for laser spectroscopy. This field was reenergized after the first experimental demonstration of electromagnetically induced transparency (EIT) in three-level atomic gas systems about twenty years ago by the group of Professor S.E. Harris at Stanford University. Such EIT effects were later observed in atomic vapor cells with low-power diode lasers under two-photon Doppler-free configurations and in cold atoms confined in magneto-optical traps, as well as in many molecular and solid material systems. In the past twenty years, many interesting phenomena related to laser-induced atomic coherence and interference were experimentally demonstrated, such as slow and superluminal light propagations, lasing without inversion, enhanced refractive index, enhanced nonlinear optical processes, matched light pulses, optical memory, correlated/entangled photon pairs, controllable optical bistability, and cavity-QED effects. The atomic coherence effects in multi-level media have been shown to have potential applications in all-optical switching/router, logic gates, precision measurements, optical buffer/delay lines for optical communication and computing, and quantum information processing.

There have been several well-written review articles published in recent years to cover various aspects of atomic coherence and interference in multi-level systems. In this e-book, we put together reviews of several research topics related to laser-induced atomic coherence and interference, and its potential applications in atomic and solid media written by active researchers working in these fields. We hope that this e-book can serve as a good reference for graduate students and researchers interested in acquiring some general understanding and perspective of this active research field.

Chapter 1 presents experimental studies of EIT-enhanced linear and nonlinear dispersions of three-level atomic medium inside an optical cavity. By balancing the sharp linear and Kerr-nonlinear dispersions of the atomic system near the EIT resonance, the total intracavity atomic dispersion can be made to be either normal or anomalous, which leads to subluminal (slow) light propagation or superluminal (fast) light propagation in the cavity, and concomitantly, results in a narrowed or broadened cavity linewidth compared to the empty cavity case. Under certain experimental conditions, the so called "white-light cavity" can be realized with a very broad transmission spectrum, which can have applications in nonlinear precision spectroscopy and recycling cavity of the laser interferometer for the gravitational-wave detection. Chapter 2 shows that by controlling the phases between various laser beams interacting with multi-level atomic systems, phase-dependent quantum interference is induced in the atomic systems and either constructive or destructive interference can be obtained in the probe transitions. For example, when bichromatic coupling and probe fields are used in a three-level system, the interference between the resonant two-photon Raman transitions can be controlled by varying the relative phases of the coupling or probe fields. Also, when two coupling fields and two probe fields are used in a four-level double-Λ system, manipulating the relative phase among the laser beams creates interference between three-photon and one-photon excitation processes, which can be used to selectively enhance or suppress probe light absorptions. Both experimental studies and simple theoretical descriptions are presented to illustrate such phase-dependent atomic coherence effects.

In Chapter 3, several schemes for realizing atomic localization via atomic coherence and quantum interference in multi-level atomic systems are proposed, including double-dark resonance effects, sub-half-wavelength localization via two standing-wave fields, and two-dimensional localization by using two orthogonal standing-wave fields. Using double-dark resonances, the detecting probability and localization precision for the atoms can be greatly enhanced. Also, using standing-wave fields in ladder-type system can improve the detecting probability and lead to sub-half-wavelength localization precision in two-dimension, which provides potential applications in two-dimensional nano-lithograph. Chapter 4 presents some recent theoretical studies on quantum correlation and entanglement properties of the four-wave mixing and quantum-beat laser systems in multi-level atomic systems. The four-wave mixing can yield Einstein-Podolsky-Rosen (EPR) entangled states by simultaneously absorbing in the excitations from a pair of squeeze-transformed modes, and the quantum-beat lasers can act as bright sources of entangled light beams with sub-Poissonian photon statistics.

Chapter 5 describes few techniques to generate and tune the photonic bandgaps of the gratings written by standing-wave fields in coherent media such as cold rubidium atoms, Pr^{3+}: Y_2SiO_4, and diamond containing N-V color centers. Both steady and dynamic optical responses of the media exhibiting induced photonic bandgaps have been theoretically examined in terms of transmitted and reflected spectra or pulse propagation dynamics. The standing-wave driving configuration is shown to be an efficient technique to control light flows and optical nonlinearities with spatially periodic quantum coherence, which may be exploited to achieve all optical switching, router, and storage and thus has potential applications in quantum information processing. Chapter 6 shows several experimental demonstrations of light storage based on atomic coherence and atomic coherence transfer. By preparing maximal atomic coherence in the coherently-driven media, light storage based on F-STIRAP is realized in atomic vapor and solid-state crystal respectively. Some applications of EIT-based light storage in solid Pr^{3+}:Y_2SiO_5 crystal are studied experimentally. Also, by employing

STIRAP process, atomic coherence transfer between different spin levels is proposed theoretically and demonstrated experimentally.

As one can see that this e-book has covered a broad spectrum of research topics from experimental demonstrations of EIT-related effects (such as cavity linewidth modification, phase-dependent interference, and optical storage) to several interesting theoretical predictions (e.g. localization of atoms, quantum correlation, and tunable photonic bandgaps). Also, in the experimental demonstrations of atomic coherence effects, different multi-level media, such as hot atomic cell in an optical cavity (Chapter 1), cold atoms in the magneto-optical trap (Chapter 2), and solid crystal at low temperature (Chapter 6), are used, which show the broad applicability of atomic coherence effects and their potential applications.

Jinyue Gao, Editor

Min Xiao, Co-Editor

Yifu Zhu, Co-Editor

CONTRIBUTORS

Haibin Wu	University of Arkansas, USA
Min Xiao	University of Arkansas, USA
Jiepeng Zhang	Florida International University, USA
Yifu Zhu	Florida International University, USA
Luling Jin	Shanghai Institute of Optics and Fine Mechanics, Chinese Academy of Sciences, Peoples Republic of China
Yueping Niu	Shanghai Institute of Optics and Fine Mechanics, Chinese Academy of Sciences, Peoples Republic of China
Hui Sun	Shanghai Institute of Optics and Fine Mechanics, Chinese Academy of Sciences, Peoples Republic of China
Shiqi Jin	Shanghai Institute of Optics and Fine Mechanics, Chinese Academy of Sciences, Peoples Republic of China
Shangqing Gong	Shanghai Institute of Optics and Fine Mechanics, Chinese Academy of Sciences, Peoples Republic of China
Guangling Cheng	Huazhong Normal University, Peoples Republic of China
Jinhua Zou	Huazhong Normal University, Peoples Republic of China
Xiaoxia Li	Huazhong Normal University, Peoples Republic of China
Jin-Hui Wu	Jilin University, Peoples Republic of China
M. Artoni	European Laboratory for Nonlinear Spectroscopy, Italy
Jin-Yue Gao	Jilin University, Peoples Republic of China
G. C. La Rocca	European Laboratory for Nonlinear Spectroscopy, Italy
Ai-Jun Li	Jilin University, Peoples Republic of China

Atomic Coherence and its Potential Applications

Chapter 1. Cavity Linewidth Controls with an Intracavity Three-level Atomic Medium

Haibin Wu and Min Xiao
University of Arkansas

Abstract: Linear absorption and dispersion properties, as well as Kerr nonlinear index, of three-level atomic systems can be greatly modified under the condition of electromagnetically induced transparency (EIT). By placing such EIT atoms in a vapor cell inside an optical ring cavity, the cavity transmission spectrum can be altered and controlled. We show that the cavity transmission linewidth can be narrowed substantially comparing to the empty cavity linewidth due to the sharp normal (linear) dispersion associated with the EIT resonance (which can be considered as photons traveling with a slower speed inside the optical cavity). On the other hand, the Kerr nonlinear dispersion has the opposite slope comparing to the linear dispersion near the EIT resonance, which can be used to balance the linear dispersion and lead to total anomalous dispersion for the intracavity atomic medium (with ``superluminal photon speed" inside the optical cavity). Such anomalous dispersion in the intracavity medium makes the cavity transmission linewidth broader than the empty cavity linewidth. Under certain parameters, the so called ``white-light cavity" condition can be satisfied, which makes the cavity transmission linewidth very broad and, at the same time, have high cavity transmission. Such modified and controlled cavity transmission linewidths can have many interesting applications in frequency locking, cavity ring-down spectroscopy, nonlinear optical spectroscopy, cavity-QED, and even recycling cavity of the gravitational-wave detector.

Introduction

It is well established that the linear dispersion can be greatly enhanced near electromagnetically induced transparency (EIT) resonance in a three-level atomic system [1]. Such a sharp normal dispersion with reduced absorption can be used to reduce the propagation of the light pulses to extremely small group velocity in the medium and thus to delay and store the optical information in the medium [2, 3]. Recently, the field of optical information storage has been unprecedentedly developed in all kinds of media. Two-entangled photon states have been mapped into two separate atomic ensembles, and after a certain time delay, those entangled states can be retrieved from the ensembles [4]. The probe optical pulse has been even stored in one Bose-Einstein condensate, and then such coherent information can be transferred into the matter wave and travel into a second condensate, which is then mapped back into its original photonic states [5]. When such EIT medium is placed inside an optical cavity, the cavity response can be drastically modified, resulting in a substantially narrowed cavity transmission linewidth [6-11]. Such cavity linewidth narrowing can be considered as due to the reduced photon group velocity propagation inside the intracavity medium, which effectively increases the photon lifetime in the cavity with a steep normal linear dispersion slope of the EIT system [7]. On the other hand, such EIT system also exhibits giant, resonantly enhanced Kerr nonlinearities due to the atomic coherence [12], which can be used to generate optical switching, quantum nondemolition measurement and quantum phase gate at the single-photon energy level. Moreover, this Kerr nonlinear dispersion typically has an opposite sign compared to the linear dispersion near the EIT resonance, i.e., a negative (or anomalous) dispersion which gives the superluminal propagation of the light pulses. In this Chapter, we review some previous experiments demonstrating that the transmission spectra can be drastically modified with such coherently-prepared atoms inside an optical ring cavity. By balancing the enhanced linear dispersion and nonlinear dispersion, the group index of the atomic medium can be easily manipulated and controlled; the cavity transmission linewidth can be made to be either far smaller or much broader than the empty cavity linewidth. With certain cavity frequency detuning, by adjusting one parameter (the cavity driving field intensity), the cavity linewidth can be made to change from far below to above the empty cavity linewidth, corresponding to the photon propagation inside the cavity from ``subluminal" to ``superluminal" speed, respectively. Under certain conditions, such negative nonlinear dispersion can be tuned to completely compensate the phase shift of the off-resonant frequency, making the round-trip phase shift basically independent of the frequency of the input light, and thus getting into the so-called "white-light cavity" (WLC) condition.

The Chapter is organized as follows. First the theoretical calculations with Λ-type three-level atoms in the optical ring cavity are described. The EIT-enhanced linear and nonlinear dispersions are calculated, and their effects to the cavity linewidth are discussed in different parametric regions. Second, various experimental observations are presented to demonstrate those interesting effects on cavity linewidth narrowing and broadening, as well as WLC. The last part serves as the summary with some discussions.

Theoretical analysis

For an optical ring cavity of length L with an intracavity medium, as shown in Fig. 1(a), the transmitted intensity I_t is given by the well-known Airy function:

$$\frac{I_t}{I_{in}} = \frac{T_1 T_2}{1 + R_1 R_2 e^{-\alpha l} - 2\sqrt{R_1 R_2 e^{-\alpha l}} \cos(\Phi)}, \qquad (1)$$

where $R_{1,2}$ and $T_{1,2}$ are the reflectivities and the transmissivities of the input and output mirrors, respectively. The mirror M_3 has nearly 100% reflectivity. α is the round-trip absorption including all losses due to the intracavity medium excluding the transmissions of input and output mirrors and Φ is the round-trip phase for cavity field.

Without the medium inside the cavity the phase shift is given by $\Phi = L\Delta / c$, where $\Delta = \omega_p - \omega_c$ is the frequency detuning from the cavity resonant frequency ω_0 and c is the speed of light in vacuum. The empty cavity linewidth [full width at half maximum (FWHM)] is

$$\Delta\omega_0 = \frac{2\pi c}{Lf}, \qquad (2)$$

where f is the cavity finesse ($f = \frac{\pi (R_1 R_2 e^{-\alpha l})^{1/4}}{1 - \sqrt{R_1 R_2 e^{-\alpha l}}}$).

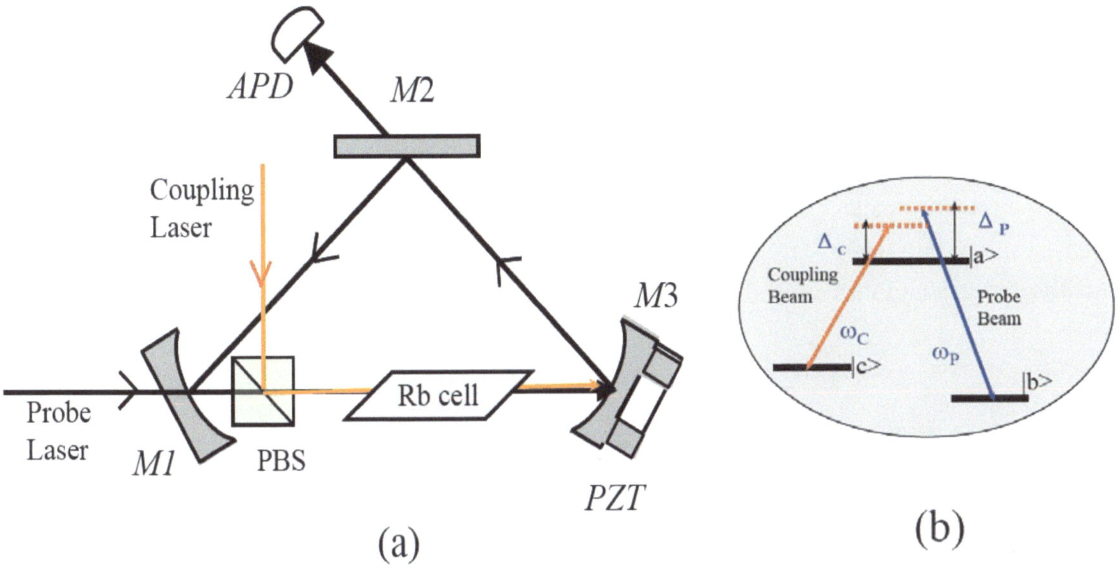

Fig. 1. Experimental setup. PBS, polarizing cubic beam splitters; M1-M3, cavity mirrors; APD, avalanche photodiode detector; PZT, piezo-electric transducer. (b) Three-level Λ-type atomic system.

If a medium of length l<L with an index of refraction n is placed inside this optical cavity, the phase shift then depends on the laser frequency and can be expressed as

$$\Phi = L\Delta / c + \frac{\omega_p l}{c}(n-1). \tag{3}$$

The refractive index n can be expanded and expressed at near the resonant frequency as

$$n = n_0 + \left.\frac{\partial n}{\partial \omega_p}\right|_{\omega_p=\omega_0} \Delta + \left.\frac{\partial^2 n}{\partial \omega_p{}^2}\right|_{\omega_p=\omega_0} \frac{\Delta^2}{2} + \left.\frac{\partial^3 n}{\partial \omega_p{}^3}\right|_{\omega_p=\omega_0} \frac{\Delta^3}{6} + O(\Delta^4), \tag{4}$$

where n_0 equals to $n(\omega_0)$.

$\cos(\Phi)$ can be expanded near the cavity resonant frequency:

$$\cos(\Phi) = 1 - \left(\frac{L}{c}\right)^2 \left[1 + \frac{l}{L}(n-1) + \omega_0 \frac{l}{L} \left.\frac{\partial n(\omega_p)}{\partial \omega_p}\right|_{\omega_p=\omega_0}\right]^2 \frac{\Delta^2}{2} + O(\Delta^3). \tag{5}$$

Drop the higher order terms, the cavity transmission intensity and the cavity linewidth can be written as

$$\frac{I_t}{I_{in}} = \frac{T_1 T_2}{1 + R_1 R_2 e^{-\alpha l} - 2\sqrt{R_1 R_2 e^{-\alpha l}} \cos\left[\frac{\Delta L}{c}\left(1 + \frac{l}{L}(n-1) + \omega_0 \frac{l}{L} \left.\frac{\partial n(\omega_p)}{\partial \omega_p}\right|_{\omega_p=\omega_0}\right)\right]} \tag{6}$$

$$\Delta\omega = \Delta\omega_0 \frac{1}{1+\frac{l}{L}(n-1)+\omega_0 \frac{l}{L}\frac{\partial n(\omega_p)}{\partial \omega_p}\Big|_{\omega_p=\omega_0}}. \tag{7}$$

When the dispersion of the medium is positive (i.e., normal dispersion), the denominator of Eq. (7) is larger than 1, which can greatly narrow the cavity linewith, as shown in Fig. 2 (neglecting the absorption). The dispersion of the atomic medium near the EIT resonant condition is very sharp

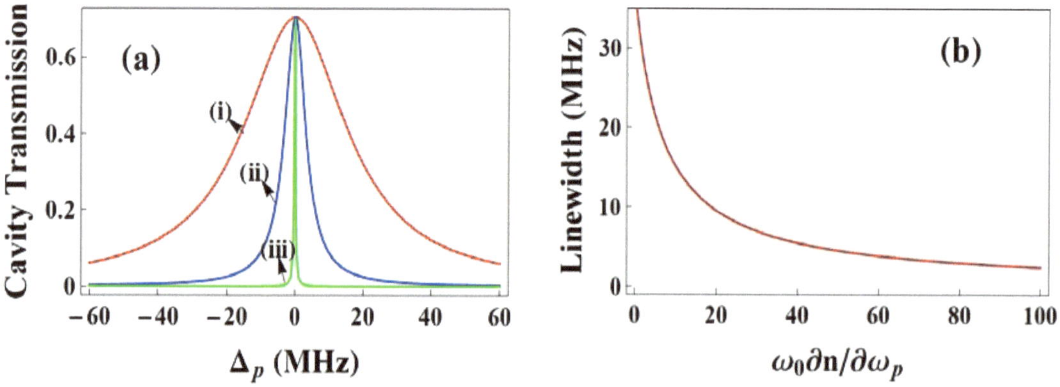

Fig. 2. (a) Cavity transmission as a function of the probe frequency detuning Δ_p (MHz) with different linear dispersions: (i) $\omega_0 \frac{\partial n}{\partial \omega_p} = 0$; (ii) $\omega_0 \frac{\partial n}{\partial \omega_p} = 10$; (iii) $\omega_0 \frac{\partial n}{\partial \omega_p} = 200$. (b) Cavity transmission linewidth as a function of $\omega_0 \partial n / \partial \omega_p$. The parameters are chosen as $R_1 = 0.97$; $R_2 = 0.986$, L=0.365 m, l=0.05 m and $\alpha = 0$.

and extremely large, so the group velocity of light pulses in the medium has been reduced down to 17 m/s [2]. Now, we consider a typical three-level atomic medium with Λ configuration as shown in the Fig. 1 (b). A probe beam with frequency ω_p near the resonant frequency ω_{ba} couples the transition $|b\rangle - |a\rangle$ and a coupling beam with the frequency $|c\rangle$ drives the transition $|c\rangle - |a\rangle$. $\Delta_p = \omega_p - \omega_{ba}$ and $\Delta_c = \omega_c - \omega_{ca}$ are the frequency detunings of the probe and coupling beams, respectively. With a perturbation method, one can solve the density-matrix equations of the system at the steady state and get the linear and Kerr nonlinear refractive indices which can be expressed as [15]

$$n_0 = 1 + \frac{iN|\mu_{ba}|^2}{2\varepsilon_0 \hbar}\frac{1}{F}, \tag{8}$$

$$n_2 = -\frac{iN|\mu_{ba}|^4}{4n_0{}^2\varepsilon_0{}^2 c\hbar^3\gamma}\frac{F+F^*}{F|F|^2}. \qquad (9)$$

Here, N is the atomic density and $F = \Delta_p - i\gamma_{ba} + \frac{\Omega_c{}^2/4}{\Delta_p - \Delta_c - i\gamma_{bc}}$, where $\Omega_c = \mu_{ca}E_c/\hbar$ is the Rabi frequency for the $|c\rangle - |a\rangle$ transition induced by the coupling field E_c. γ_{bc} is the ground-state decoherence rate and γ_{ba} is the spontaneous decay rate from $|a\rangle - |b\rangle$. μ_{ba} is the dipole moment of the probe transition $|b\rangle - |a\rangle$.

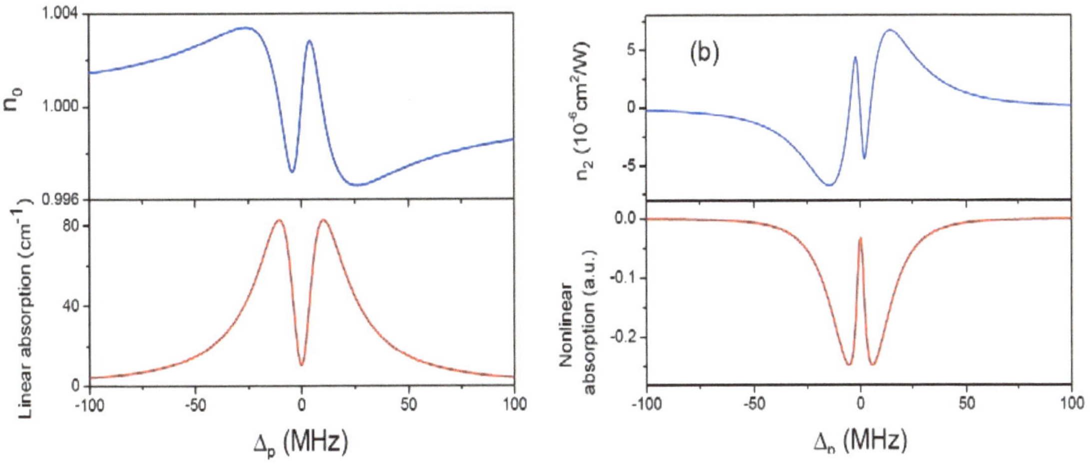

Fig. 3. (a) linear and (b) nonlinear refractive indices and their corresponding absorptions as a function of Δ_p, respectively. The parameters used in the calculations are: $\gamma_{ba} = 2\pi \times 3$ MHz, $\gamma_{ba} = 2\pi \times 0.1$ MHz, $\Delta_c = 0$, $\Omega_c = 2\pi \times 4$ MHz and $N = 5\times 10^{11}/cm^3$.

The linear and nonlinear Kerr refractive indices, as well as their corresponding absorptions, are shown in Fig. 3 as a function of the probe frequency detuning Δ_p. We can clearly see that near the EIT resonance both the linear and nonlinear refractive indices have very sharp slopes but with opposite signs. Figure 4 shows the derivatives of the linear and nonlinear refractive indices as a function of the probe frequency detuning [16]. At small probe frequency detuing, the linear dispersion is normal and positive while the nonlinear dispersion is anomalous and negative. This sensitive dependence on probe frequency detuning can be used to obtain the desired dispersion values for certain applications. Figure 5 depicts $\frac{\partial(n-n_0)/\partial\omega_p}{\partial n_0/\partial\omega_p}$ as a function of the probe

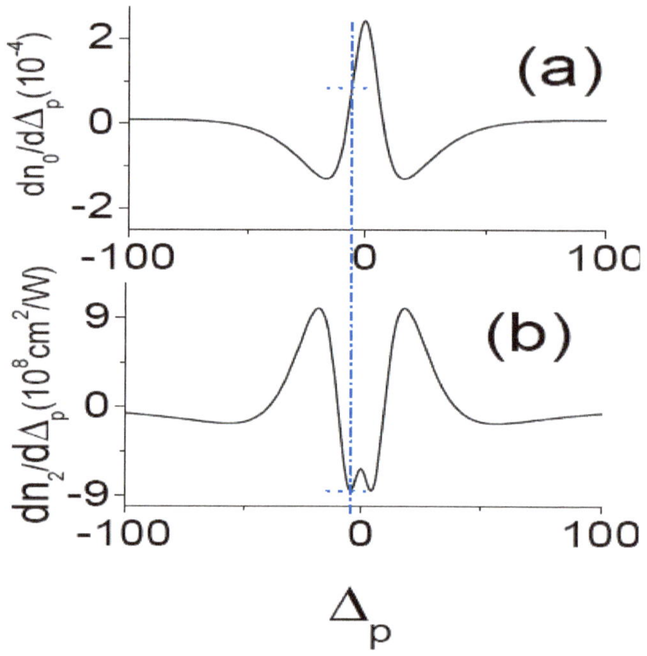

Fig. 4. The derivatives of the linear (a) and nonlinear (b) refractive indices as a function of Δ_p, respectively. The parameters used here are the same as Fig. 3. Adopted from Ref. [16].

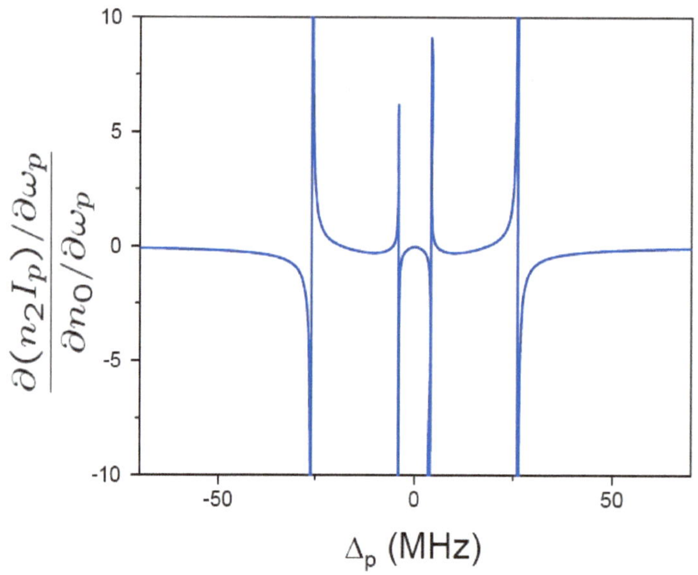

Fig. 5. $\dfrac{\partial(n_2 I_p)/\partial\omega_p}{\partial n_0/\partial\omega_p}$ as a function of the probe frequency detuning Δ_p. Here the intensity of the probe beam is chosen as $I_p = 100\ W/cm^2$, other parameters are the same as in Fig. 3.

frequency detuning Δ_p. By adjusting the input intensity and the probe frequency detuning Δ_p, one can easily control the nonlinear dispersion and linear dispersion, as well as their derivatives, to get any desired values of n_g. The group index n_g is defined as [13] $n_g = n + \omega_p \frac{\partial n}{\partial \omega_p}$ and for a Kerr nonlinear medium with $n = n_0 + n_2 I$, it can be written as [16]

$$n_g = n_0 + n_2 I_p + \omega_p \left(\frac{\partial n_0}{\partial \omega_p} + \frac{\partial n_2}{\partial \omega_p} I_p \right). \tag{10}$$

When the intensity of the probe beam (I_p) is low, the linear normal dispersion term dominates, so $n_g \gg 1$ and the group velocity of the light pulses can be slowed down; as the intensity of the input light increases, the nonlinear dispersion term becomes important, n_g can be tuned to be zero and even negative (since $\frac{\partial n_0}{\partial \omega_p} > 0$ and $\frac{\partial n_2}{\partial \omega_p} < 0$ near the EIT resonance), thus making the superluminal propagation of the light pulses (or photons) [13] due to the total negative dispersion of the EIT medium.

When such coherently prepared atomic medium is placed inside the optical cavity, the modified cavity linewidth can be written as [16]

$$(\Delta \nu) = \frac{(\Delta \nu_0)(1-\sqrt{R_1 R_2 \kappa})}{\sqrt{\kappa}(1-\sqrt{R_1 R_2})} \frac{1}{1+\frac{l}{L}(n_g - 1)} = (\Delta \nu)_0^{'} \frac{1}{1+\frac{l}{L}(n_g - 1)}, \tag{11}$$

where $(\Delta \nu)_0$ is the empty cavity linewidth and $(\Delta \nu)_0^{'}$ is the modified empty cavity linewidth including all kinds of absorption elements in the cavity; R_1 and R_2 are the reflectivities of the cavity input and output mirrors, respectively; $\kappa = e^{-\alpha l}$ is the single-pass absorption of the intracavity medium; l and L are the lengths of the medium and cavity, respectively. Without considering the nonlinear contributions, Eq. (11) reduces to the result of the linear case in which the cavity linewidth is substantially narrowed due to $n_g \gg 1$. In linear case, since the dispersion is always normal near the EIT resonance, only cavity linewidth narrowing can be obtained [7]. When the intensity of the cavity field increases, the negative nonlinear dispersion term in Eq. (10) begins to play an important role, which can make the group refractive index change from $n_g > 1$ to $n_g < 1$, corresponding to the cavity transmission linewidth change from below to above the empty-cavity linewidth. Such cavity linewidth narrowing and broadening due to the dispersive intracavity medium are the manifestations of the "slow" (subluminal) and "fast" (superluminal) light propagations inside the optical cavity, respectively.

Checking Eq. (11), not only the cavity transmission linewidth can be larger than the empty cavity linewidth (when $n_g < 1$) but it can also become infinite when $1 + l / L(n_g - 1) = 0$ (actually it cannot be infinite due to the contributions of the higher-order dispersions which have neglected here). This condition is n_g if the medium fills the cavity (l=L). Otherwise, it is given in general by [17]

$$n_g = 1 - \frac{L}{l}, \qquad (12)$$

which is the so-called WLC condition [18]. Under this condition the nonlinear dispersion compensates the phase shift of the off-resonant frequency and thus makes it independent of the frequency of the input light beam. So the cavity linewidth becomes very broad, as shown in Fig. 6. Such broad cavity transmission comes from the compensation of phase shift of the off- frequency of the cavity input light, which is easily understand with following formula. The round-trip phase shift of intracavity field can be write as $\Phi = 2\pi v L / c = 2\pi m + 2\pi \Delta v L / c + 2\pi v \Delta L / c$. Under WLC condition, $\Phi = 2\pi m + 2\pi v \Delta L / c$, although the round-trip shift depends on the cavity length, it is independent of the frequency of the input light, making the cavity transmission very broad.

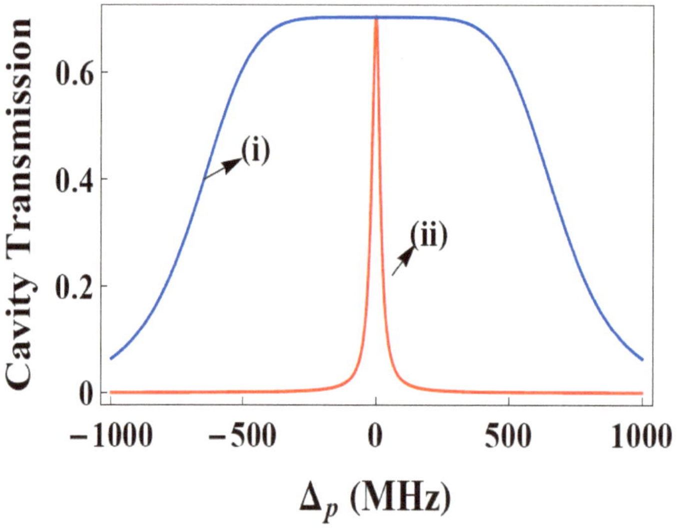

Fig. 6. Cavity transmission as a function of the probe detuning Δ_p (MHz). (i) WLC transmission; (ii) the empty cavity transmission. Here, the intensity of the probe beam is $I_p = 20 \quad W / cm^2$ and the cavity frequency detuning is $\Delta_\theta = 0$, other parameters are the same as in Fig. 2.

Since the linear and Kerr nonlinear dispersions in such EIT medium change dramatically as functions of various parameters, such as the coupling beam Rabi frequency and its frequency detuning, the probe beam frequency detuning, and the cavity frequency detuning, the WLC condition can be easily satisfied by choosing the appropriate and different sets of parameters. To better illustrate this flexibility, Fig. 7 plots n_g as functions of the coupling Rabi frequency and

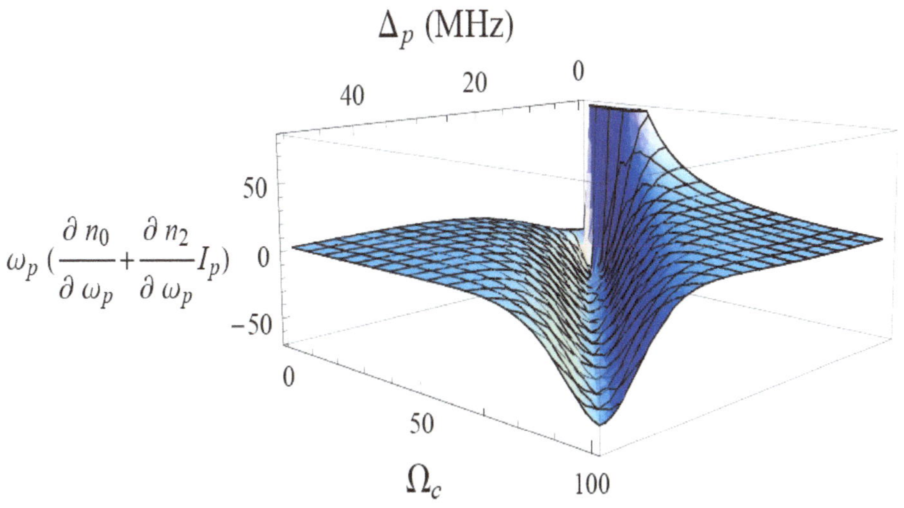

the probe beam frequency detuing Δ_p: for a given (arbitrary) cavity input intensity, by adjusting Ω_c and Δ_p , $n_g = 1 - L/l$ can always be satisfied.

Fig. 7. n_g as functions of the probe beam frequency detuning Δ_p and the coupling beam Rabi frequency Ω_c. Here the intensity of the probe beam is $I_p = 20W/cm^2$, other parameters are the same as in Fig. 3.

The above calculations show that the interplays between the greatly enhanced linear [1] and nonlinear dispersions [15] in an EIT atomic medium can significantly modify the cavity output spectrum, when it is placed inside an optical cavity. Interesting phenomena, such as cavity linewidth narrowing, broadening, and WLC, can be expected in such system involving EIT atoms and an optical cavity. Although the above calculations are done without considering the Doppler effect, and the experiments are carried out in a system involving atomic vapor cell, the predicted phenomena can still be observed by employing the two-photon Doppler-free configuration with the two laser beams propagating through the three-level Λ-type atomic medium co-linearly, such eliminating the first-order Doppler effect [19]. The residual Doppler effect can be easily taken into account by integrating the expressions over the atomic velocity distribution, as done in Refs. [1, 19].

Experimental setup and observations

The experiments are done in a system with a 5 cm long rubidium atomic vapor cell placed inside a 37 cm long three-mirror ring cavity, as shown in Fig. 1 (a). D_1 lines of ^{87}Rb atoms are used to the probe field to interact with the transition of $5S_{1/2}$, F=1 $\rightarrow 5P_{1/2}$, F=2. The coupling beam couples the transition of $5S_{1/2}$, F=2 $\rightarrow 5P_{1/2}$, F=2, as shown in Fig. 1 (b). The coupling laser beam enters the cavity through PBS and does not circulate in the cavity. The coupling and probe beams co-propagate through the Rb vapour cell in order to effectively eliminate the first-order Doppler effect near the two-photon resonance [19], and they are orthogonally polarized to maximize the interactions. The radii of the coupling and probe beams at the center of the cell are estimated to be 600 μm and 100 μm, respectively. The ``empty" cavity finesse modified by the minor losses due to the intracavity PBS and vapor cell windows is about 48 (with the frequency of the probe beam far off atomic resonances), corresponding to a cavity transmission linewidth of 17 MHz (The finesse measured is about 100 and the linewidth is only about 5.75 MHz with no intracavity elements). The atomic cell has two Brewster windows and is wrapped inside the μ-metal sheets to protect from the disturbance of the stray magnetic fields. The transmissivities of the input mirror M_1 and output mirror M_2 are 3.0 % and 1.4 %, respectively. The third mirror M_3 is a high reflector and mounted on a PZT for cavity length scanning and locking.

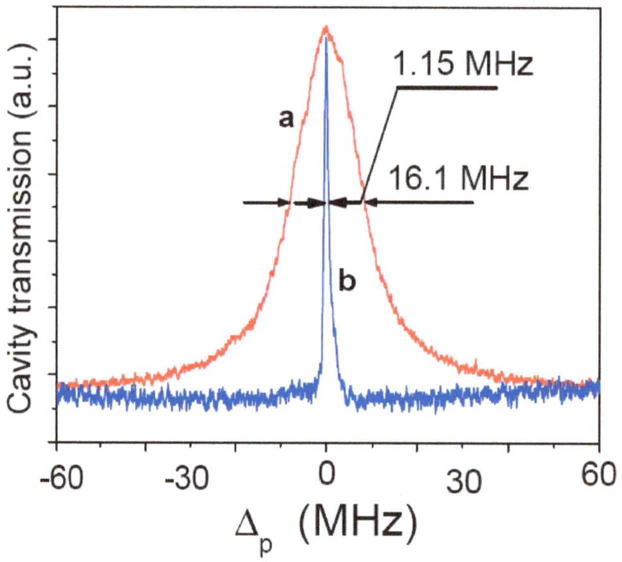

Fig. 8. Cavity transmission versus probe frequency detuning for: (a) No coupling beam and the probe frequency is well outside the absorption line, and (b) Coupling beam power of 0.81 mW and the probe frequency is scanned through the probe transition ω_{ab}. Both curves: $\Delta_c = 0$, $T = 87^0C$. Note: the intensity scales for (a) and (b) are different. Adopted from Ref. [7].

At the weak-field driving condition, the linear dispersion of the intracavity EIT medium dominates; the cavity transmission linewidths can be significantly narrowed. Figure 8 shows a comparison of the cavity transmission linewidth without (empty cavity) and with the coupling beam interacting with the intracavity atomic medium [7]. The experimental parameters are: the coupling laser the cavity transmission linewidth without (empty cavity) and with the coupling beam interacting with the intracavity atomic medium [7]. The experimental parameters are: the coupling laser power $P_c = 0.81$ mW, $T = 81^0 C$, $\Delta_c = 0$ and cavity frequency detuning $\Delta_\theta = 0$. The ``empty " cavity (containing the Rb vapor cell and the beam splitter but with laser frequency tuned from atomic resonance) linewidth is about 16.1 MHz. With the coupling beam on and the probe beam near the atomic resonant transition, the linewidth is changed into about 1 MHz, which is about a factor of 14 narrower than the ``empty" cavity linewidth. The dependence of the cavity linewidth on the intensity of the coupling beam is shown in Figure 9 [7]. For very low coupling powers the cavity linewidth is broad because the EIT and sharp dispersion will not be fully developed and the two-level atomic absorption makes the linewidth broad. As the coupling power increases, the cavity linewidth narrows rapidly until it reaches a minimum and then slowly broadens. The minimum cavity transmission linewidth achieved here is limited by the linewidths

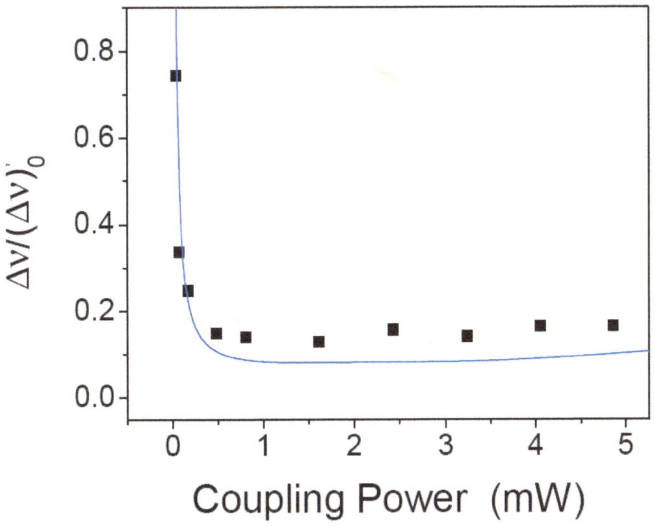

of the diode lasers which are about 1 MHz themselves. The increasing cavity linewidth at higher couplping power is due to power broadening. The solid line is a theoretical plot from Eq. (11) with integration to the velocity distribution [7].

Fig. 9. Ratio of EIT-narrowed linewidth $\Delta\nu$ to the empty cavity linewidth $(\Delta\nu)_0$ measured as a function of the coupling power for T = 87 ^{0}C . The solid curve is the fitting curve with the parameters: $\gamma = 2\pi \times 4.0$ MHz, $\gamma_{31} = 2\pi \times 3.2$ MHz. Adopted from Ref. [7].

When the cavity driving field intensity increases, then the nonlinearity becomes important, the cavity transmission shows different behaviors. Figure 10 presents the measured two cavity transmission spectra with different input intensities (or power P_P) [16]. The experimental parameters are: coupling power P_c=24 mW, coupling beam frequency detuning $\Delta_c = 0$, cavity detuning $\Delta_\theta = -40$ MHz and the cell temperature $T = 70^0 C$. The inset of Fig. 10 gives the empty cavity linewidth of $(\Delta\nu)_0 = 17$ MHz.

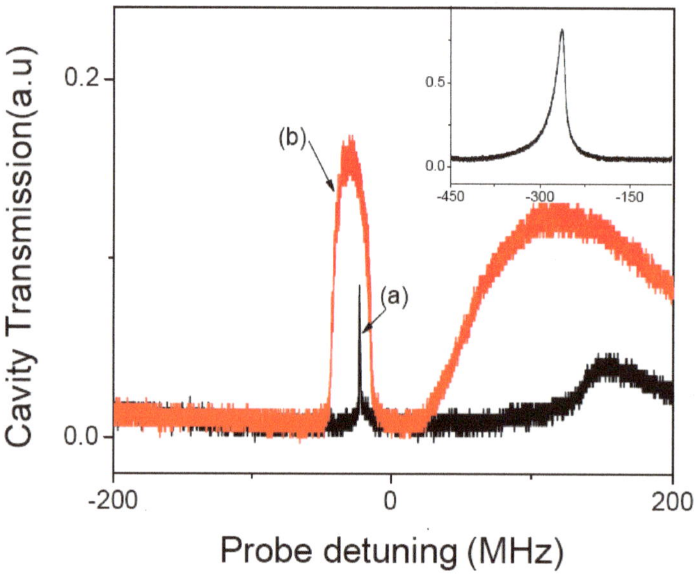

Fig. 10. Two typical cavity transmissions with different input intensities. (a) Cavity transmission at $P_p = 2.7$ mW; (b) cavity transmission at $P_p = 7.4$ mW. Inset: empty cavity transmission (the laser frequency is tuned far from atomic transition). Other experimental parameters are: $P_c = 24$ mW, T=70 ^{0}C, $\Delta_c = 0$, $\Delta_\theta = -60$ MHz. Adopted from Ref. [16].

For $P_p = 2.7$ mW, the measured linewidth is about $(\Delta\nu) = 0.9$ MHz by which the group index is estimated to be $n_g = 133$, as shown in the curve (a). In this case the cavity linewidth is significantly narrowed, corresponding to the ``slow" light propagation in the cavity. For the cavity input power $P_p = 7.4$ mW, nonlinear dispersion becomes dominant and broadens the cavity transmission with a measured linewidth of about $(\Delta\nu) = 28$ MHz, which is about two times broader than the ``empty" cavity linewidth, as shown in the curve (b). Figure 11 gives the dependence of the cavity linewidth on the coupling beam power for two different probe beam powers, $P_p = 2.7$ mW and $P_p = 7.4$ mW, corresponding to the $n_g > 1$ (slow light) and $n_g < 1$ (superluminal) regions, respectively. As the coupling beam power decreases, these two very different linewidths move towards the empty cavity linewidth value of 17 MHz, which is expected since both the linear and nonlinear dispersion slopes get flat at small coupling beam power. At very high coupling beam power, the cavity linewidth will not change much due to power broadening.

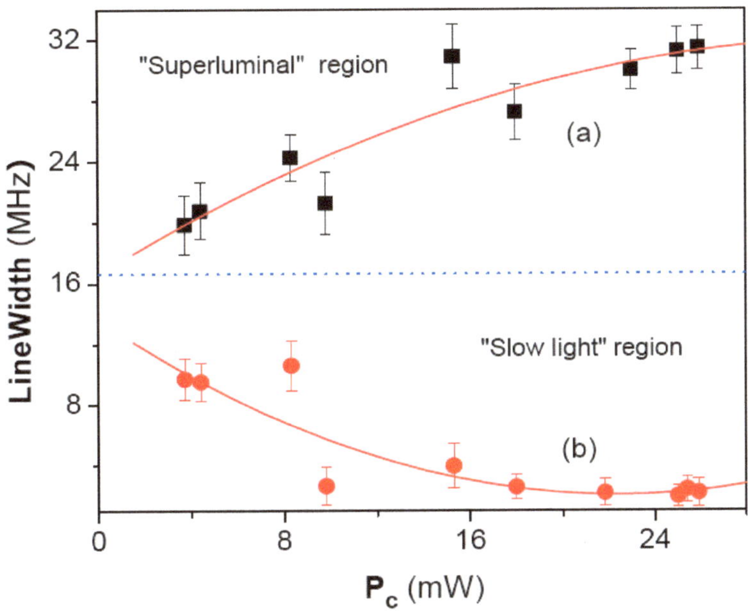

Fig. 11. Experimentally measured cavity linewidth as a function of the coupling laser power. The curve (a) is for $P_p = 7.4$ mW (in the superluminal region) and curve (b) for $P_p = 2.7$ mW (in the slow light region). Other experimental parameters are same as in Fig. 10. The solid curves are best fits of the data using quadratic function. Adopted from Ref. [16].

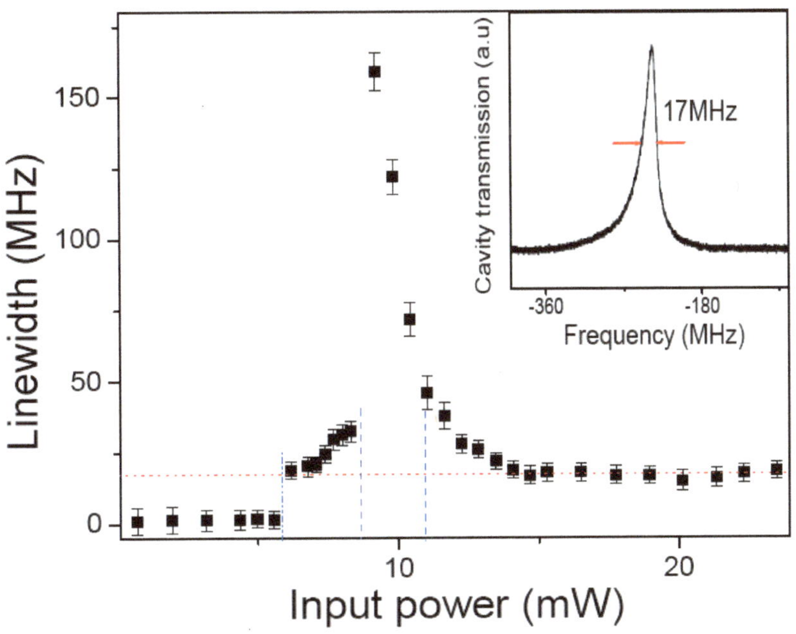

Fig. 12. Experimentally measured cavity linewidth versus the cavity input power. Dotted horizontal line corresponds to the empty cavity linewidth; Squares are the linewidths measured. Inset: ``empty"

cavity transmission (at far from atomic transition). The experimental parameters are: $P_c = 24$ mW, coupling beam frequency detuning $\Delta_c = 0$, and the cavity detuning T=70 ^{0}C, $\Delta_\theta = 0$ MHz. Adopted from Ref. [17].

Figure 12 shows the measured cavity transmission linewidth as a function of the cavity input power P_p (which is proportional to I_p). At the low cavity input powers ($P_p < 5.6$ mW), the cavity linewidth is less than the empty cavity linewith and is approximately 1MHz as in the case of the linear case [7]. As the cavity input power increases to about $P_p < 5.6$ mW (corresponding to $n_g \approx 1$), a sharp turning point appears, where the cavity linewidth jumps abruptly from narrower to broader than the empty cavity linewidth ($\Delta\nu = 17$ MHz). This behavior can be easily understood as follows. The derivative of the cavity linewidth relative to the cavity input power near the point of $n_g = 1$ is given by

$$\frac{\partial(\Delta\nu)}{\partial P_p} \approx -(\Delta\nu)_0' \frac{l}{V}\frac{\partial n_2}{\partial \omega_p}, \qquad (13)$$

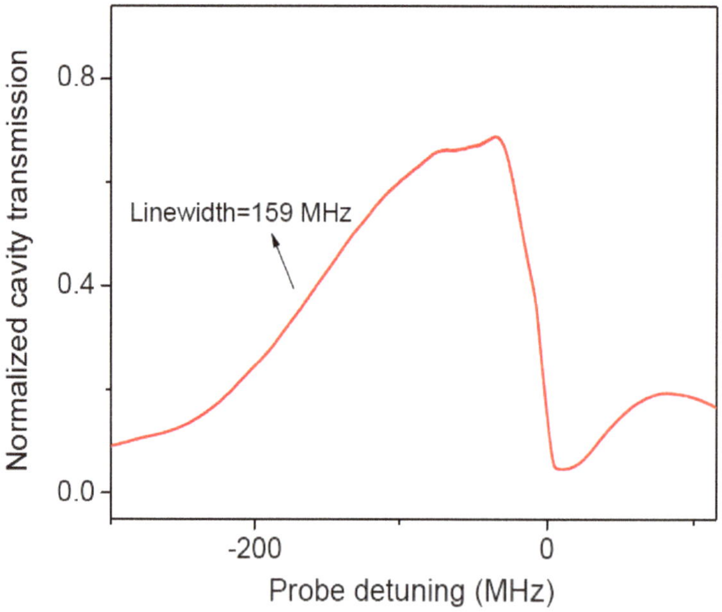

Fig. 13. The typical cavity transmission profile when the white-light cavity condition is satisfied. Cavity input power: 9.2 mW. Other experimental parameters are same as Fig. 12. Adopted from Ref. [17].

where V is the cavity mode volume. Since $\omega_p \frac{\partial n_2}{\partial \omega_p}$ is very large (much bigger than n_2) and is negative at $n_g = 1$ (which can be seen from Fig. 3), this slope can be very large and positive. In such case, the linewidth should be changed suddenly at this ``critical" point as the cavity input power changes acrossing this "critical" point. When the cavity input power increases to about $P_p = 8.3$ mW, the WLC condition (Eq. (12)) is basically satisfied (when n_g is about -6 in the experiment). The cavity transmission becomes very broad as shown in Fig. 13 which is predicted by Eq. (11) and shown in Fig. 6. The cavity linewidth measured in the experiment reaches 159 MHz, which is about ten times larger than the ``empty" cavity linewidth. It is worth to note that

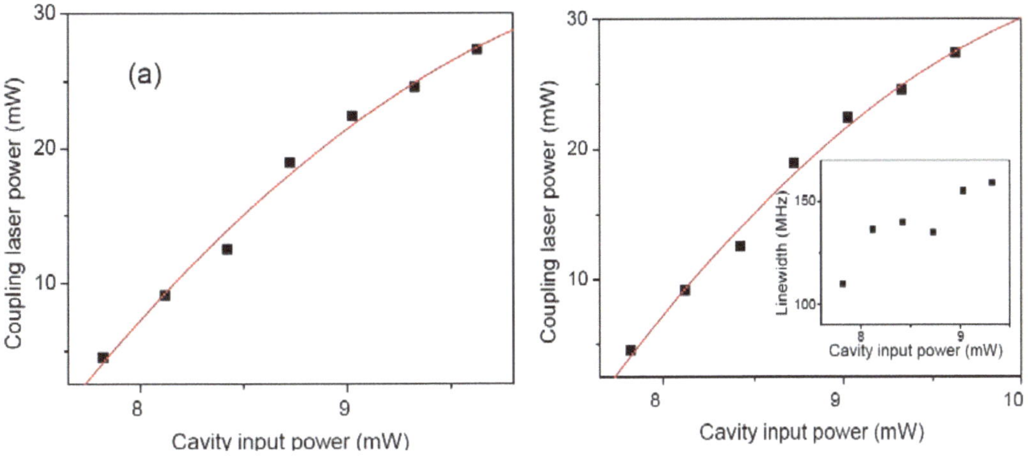

the cavity transmission is still quite high with such a broad transmission (about 70 % of the empty cavity transmission) due to the unique EIT feature and the large nonlinearity under this condition. As P_p increases further, (Δv) goes back to the empty cavity linewidth and keeps there.

Fig. 14. (a)Measured cavity linewidth as a function of coupling laser power at the "critical" intensity (the cavity input intensity satisfying WLC condition). (b) The appropriate coupling laser power for the cavity input power when the white-light condition is satisfied. Inset: the corresponding linewidths measured for each point on Fig. 14 (b). Solid line is the best fit with a quadratic function. Adopted from Ref. [17].

When the conditions reach the WLC, one can fix the cavity input intensity and measure the cavity transmission linewidth, the experimental results are shown in Fig. 14 (a). It is clear that the linewidth can be easily controlled with the coupling laser beam power. Another strength with the current mechanism is that one has more freedom to choose the parameters to satisfy the WLC condition in the parameters space as shown in Fig. 7. For a given cavity input intensity, one can always find a set of parameters to balance the linear and nonlinear refractive indices, as well as their derivatives, to reach the WLC condition and make the cavity transmission linewidth maximum. Figure 14 (b) just presents a cross section of the available entire parameter space (such as represented in Fig. 7) in which for each given cavity input intensity (power) a coupling power can always be found to satisfy Eq. (12), i.e., WLC condition, the corresponding (optimized) linewidth is shown in the inset of Fig. 14 (b).

This feature can be very useful towards the applications which are not very sensitive to the fluctuation of the input intensity. We have also obtained very broad transmission spectra in the experiment with adjusting other parameters such as the cavity and coupling laser frequency detunings.

Such WLC phenomenon has also been observed in the system with Raman gain system [20] and recently theoretically predicted in a four-wave mixing system [21].

The WLC can be used to increase the sensitivity of the recycling cavity of the Advanced Gravitational-wave Detectors such as LIGO, in which both light transmission and broad spectral response are needed. The technique described here with balancing the linear and Kerr nonlinear dispersions will be particularly useful when the power in the signal recycling cavity becomes large. Also, this WLC can be used for nonlinear spectroscopy, pulse shaping, and laser cooling of the atoms and ions [22-24].

An application of large linear dispersion: cavity ringdown effect

When the scan speed of an optical cavity length is faster than the round-trip speed of the intracavity photon, the cavity output field profile shows an amplitude oscillation on the background of the exponential decay, which is known as the cavity ringdown effect (CRE) [25, 26]. Such an oscillation comes from the interference of the original input laser field and the intracavity field. The oscillation period is determined by the cavity scan speed. Although such CRE is very useful in many applications especially in measuring small traces of chemical and biological systems [25, 26], it has several serious limitations i.e., it needs to have the high cavity finesse, long cavity length and fast cavity length scan. However, by using the highly dispersive intracavity medium, the conditions to observe the CRE can be greatly relaxed, such as employing the great normal dispersion near the two-photon resonance with the coherent EIT medium in the cavity [27].

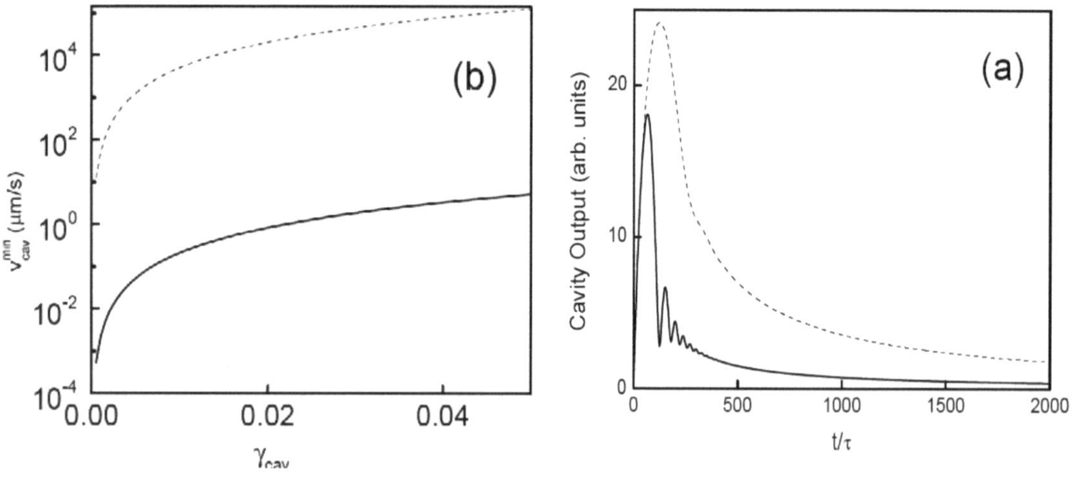

Fig. 15. (a) The transmission with the empty cavity and EIT medium inside the cavity. (b) The required minimum cavity scan speed as a function of cavity decay. The black dashed line is for an empty cavity and the blue solid line is for the cavity with an intracavity EIT medium. The parameters are used: $\kappa = 0.02$, $T = 0.03$, the scan speed v_{cav} are $2\times10^4 \mu m/s$ and $4\mu m/s$ for an empty cavity

and with an EIT medium in the cavity, respectively; and $\eta = 24000$. Adopted from Ref. [27].

Consider the system with a coherent EIT medium inside an optical cavity, as shown in Fig. 1(a). After propagating through the medium the cavity field amplitude E_p experiences a phase change which is multiplied by the factor $\exp(i\Phi)$. Under the small-gain approximation, this exponential may be expanded to the first order, and the rate of change of the cavity field due to the medium can be approximated by E_p / τ_0,

where τ_0 is the round-trip time for the photons in the cavity. The dynamical equation of the intracavity field amplitude can be written as

$$\tau_0 \frac{\partial E_p}{\partial t} = \sqrt{T_1} E_p^{in} + i\Phi_{cav} E_p - \kappa E_p, \tag{14}$$

where κ is the cavity loss; E_p^{in} is the driving term of the cavity field.

For an empty cavity the round-trip phase shift is just determined by the length of the cavity and can by expressed as $\Phi_{cav} = k(L_0 + vt)$, where k is the wave number of the cavity field and v is the cavity scan speed. So the intracavity field can be analytically solved to give

$$E_p = \frac{\sqrt{\pi T_1 E_p^{in}}}{2\sqrt{kv\tau_0}} \exp\left[\frac{i(kvt + i\kappa)^2}{2kv\tau_0}\right]\left[(-1)^{3/4}\sqrt{2} erf\left(\frac{(1+i)(kvt+i\kappa)}{2\sqrt{kv\tau_0}}\right) + (1+i) erfi\left(\frac{\kappa(1+i)}{2\sqrt{kv\tau_0}}\right)\right] \tag{15}$$

where erf is the error function. With certain parameters, the cavity transmission profile shows the oscillation, but the scan speed of the cavity length must be fast and the cavity finesse must be high. As shown in Fig. 2 (a), the cavity transmission linewidth can be greatly narrowed due to the large normal dispersion when an EIT medium is placed inside the cavity, which means that the cavity decay rate becomes smaller. The photon round-trip time is then given by $\tau_0(1+\eta)$, where $\eta = \omega_0 (l / L_0)(\partial n_0 / \partial \omega_p)|_{\omega_p = \omega_0}$. With large linear normal dispersion, $\omega_p \frac{\partial n_0}{\partial \omega_p}$ is very large and positive, leading to a greatly increase in the photon life time in the cavity, or the cavity finesse. The equation governing the evolution of the cavity field becomes

$$\tau_0 \frac{\partial E_p}{\partial t} = \frac{\sqrt{T_1}}{1+\eta} E_p^{in} + i\frac{kvt}{1+\eta} E_p - \kappa^{EIT} E_p \tag{16}$$

In such case, the effective cavity finesse is increased by a factor of $(1+\eta)$. Thus, the required cavity scan speed to observe CRE for a cavity with an intracavity EIT medium can be much slower than that for an empty cavity by a factor of $(1+\eta)$. The figure 15(a) shows the cavity transmission profiles with the normalized time t/τ. The black dashed curve is the cavity transmission without the EIT medium inside the cavity. The blue solid curve is the cavity transmission with an intracavity EIT medium. The used parameters are: $\kappa = 0.02$, $T = 0.03$. The scan speeds (v_{cav}) are $2\times10^4 \mu m/s$ and $4\mu m/s$ for an empty and with an EIT medium in the cavity, respectively; and $\eta = 24000$. It is clear that, with the help of a large linear normal dispersion of the EIT medium, the CRE can be observed for cavities with much lower finesse and with much slower cavity scaning speeds. Fig. 15 (b) shows a comparison of the minimum cavity scan speed required for observing the CRE with or without the EIT medium inside the cavity. One can see that the cavity scan speed can be significantly reduced in the case with EIT medium in the cavity, which makes the CRE easier for practical applications.

Discuss and summary

Although the sharp linear dispersion near the EIT resonance [1] has been used widely to slow down the group velocity of light pulses [3, 5], optical storage [3], and altering the cavity transmissions [6-11], the enhanced nonlinear dispersion in such EIT medium [12] is seldom considered as an useful tool to control the atom-cavity interactions. In the past, the nonlinear dispersion has been viewed as a limiting factor to achieve large group velocity reduction, since it provides a counter effect (with an opposite sign) to the large normal linear dispersion. Here, we have shown that the large Kerr nonlinear (anomalous) dispersion near the EIT resonance due to atomic coherence can actually be used to totally balance the linear (normal) dispersion at high cavity field intensities [16, 17]. Such the total intracavity dispersion for EIT atoms inside an optical cavity can be tuned from large normal dispersion (with linear dispersion dominant0 to large anomalous dispersion (with dominated nonlinear dispersion). Two interesting effects occur at the points when the group index $n_g = 1$ and $n_g = 1 - L/l$, respectively. When the cavity input intensity increases, the group velocity of the intracavity photons change from subluminal ($n_g > 1$) to superluminal ($n_g < 1$), showing a sharp change in the cavity linewidth (Fig. 12). It will be interesting to consider what happens at the exact point of $n_g = 1$. What will be the group velocity of the photons at such condition? When $n_g = 1 - L/l$, the negative (anomalous) nonlinear dispersion is large enough to compensate both the positive linear (normal) dispersion and the round-trip phase shifts for different frequencies, so the cavity transmission will be independent of the probe frequency, giving a ``white-light cavity" condition. In this case, not only the cavity transmission spectrum can be very broad (much more than the empty cavity linewidth), but also the transmission loss is small, which is very important in many applications. The exact behaviors of the cavity transmission near the WLC condition depend on the higher-order nonlinear dispersions and losses in the intracavity medium, which have been neglected in our simple theoretical treatment.

Also, the residual Doppler effect has not been considered in the theoretical calculations for the nonlinear dispersion. Although two-photon Doppler-free configuration [19] has been used in all the experiments described in this Chapter, it can only cancel the first-order Doppler effect. To consider the residual Doppler effect due to the ground-state frequency difference, integration over the velocity distribution for the hot atoms should be carried out [19], which was not done in our simple treatments [16, 17].

In summary, we have experimentally demonstrated cavity linewidth narrowing, broadening, and WLC in the three-level EIT atoms inside an optical ring cavity. By making use of the largely enhanced linear and Kerr nonlinear dispersions the total intracavity dispersion can be tuned easily from positive (normal) to negtive (anomalous) by simply increasing the probe beam intensity. Such balancing of the linear and nonlinear dispersions can give precise control over the total intracavity dispersion and reach critical points where the linear and nonlinear dispersions either cancel each other ($n_g = 1$) or the total cavity dispersion (including the round-trip phase shift without the medium due to propagation) diminishes (WLC condition). Cavity linewidth narrowing was shown to be down to 0.9 MHz (limited by the laser linewidths), and at the WLC condition, the cavity linewidth broadening was shown to be up to 10 times of the empty cavity linewidth. All this large cavity linewidth tuning range can be achieved by adjusting one parameter-input probe laser power. It was also shown that WLC condition can be satisfiedm in different sets of experimental parameters. Finally, a simple theoretical example was given to use the largely-enhanced linear dispersion in intracavity EIT medium to improve the cavity-ring down effect to make it easier for practical applications.

We acknowledge the funding support from the National Science Foundation.

References

1. Xiao M, Li Y, Jin S and et al. Measurement of dispersive properties of electromagnetically induced transparency in rubidium atoms. Physical Review Letters 1995 Jan 30;74(5):666-669.
2. Hau LV, Harris SE, Dutton Z and et al. Light speed reduction to 17 metres per second in an ultracold atomic gas. Nature 1999 Feb 18;397:594-598.
3. Phillips DF, Fleischhauer A, Mair A and et al. Storage of light in atomic vapor. Physical Review Letters 2001 Jan 29;86(5):783-786.
4. Choi KS, Deng H, Laurat J and et al. Mapping photonic entanglement into and out of a quantum memory. Nature 2008 Mar 6;452:67-71.
5. Ginsberg NS, Garner SR, and Hau LV. Coherent control of optical information with matter wave dynamics. Nature 2007 Feb 8;445:623-626.
6. Lukin MD, Fleischauer M, and Scully MO. Intracavity electromagnetically induced transparency. Optics Letters 1998 Feb 15;23(4):295-297.
7. Wang H, Goorskey D, Burkett WH and et al. Cavity-linewidth narrowing by means of electromagnetically induced transparency. Optics Letters 2000 Dec 1;25(23):1732-1734.

8. Muller G, Muller M, Wicht A and et al. Optical resonator with steep internal dispersion. Physical Review A 1997 Sep;56(3):2385-2389.

9. Soljacic M, Lidorikis E, Hau LV and et al. Enhancement of microcavity lifetimes using highly dispersive materials. Physical Review E 2005 Feb 8;71(2):026602(5).

10. Li YH, Jiang HT, He L and et al . Linewidth narrowing in microstrip resonator using effective highly dispersive medium. Chinese Physics Letters 2007 Apr;24(4):975-978.

11.Yannick Dumeige, téphane Trebaol, and Patrice Féron. Intracavity coupled-active-resonator-induced dispersion. Physical Review A 2009 Jan 30;79(1):013832(10).

12. Wang H, Goorskey D, and Xiao M. Enhanced Kerr nonlinearity via atomic coherence in a three-level atomic system. Physical Review Letters 2001Aug 13;87(7):073601(4).

13. Boyd WR, Gauthier JD. ``Slow'' and ``fast'' light. Progress in Optics 2002;43:497-530.

14. Lukin MD and Imamoglu A. Nonlinear optics and quantum entanglement of ultraslow single photons. Physical Review Letters 2000 Feb 14;84(7):1419-1422.

15. Wang H, Goorskey D, and Xiao M. Atomic coherence induced Kerr nonlinearity enhancement in Rb vapour. Journal of Modern Optics 2002 Mar 10;49(3-4):335-347.

16. Wu H and Xiao M. Cavity linewidth narrowing and broadening due to competing linear and nonlinear dispersions. Optics Letters 2007 Nov 1;32(21):3122-3124.

17. Wu H and Xiao M. White-light cavity with competing linear and nonlinear dispersions. Physical Review A 2008 Mar 6;77(3):031801(4).

18. Wicht A, Danzmann K, Fleischhauer M and et al. White-light cavities, atomic phase coherence, and gravitational wave detectors. Optics Communications 1997 Jan 15;134:431-439.

19. Gea-banacloche J, Li Y, Jin S et al. Electromagnetically induced transparency in ladder-type inhomogeneously broadened media: Theory and experiment. Physical Review A 1995 Jan;51(1):576-584.

20. Pati GS, Salit M, Salit K and et al. Demonstration of a tunable-bandwidth white-light interferometer using anomalous dispersion in atomic vapor. Physical Review Letters 2007 Sep 27;99(13):133601(4).

21. Fleischhaker R, and Evers J. Four-wave mixing enhanced white-light cavity. Physical Review A 2008 Nov 17;78(5):051802 (4).

22. Zhu M, Oates CW, and Hall JL. Continuous high-flux monovelocity atomic beam based on a broadband laser-cooling technique. Physical Review Letters 1991 Jul 1;67(1):46-49.

23. Parkins AS and Zoller P. Laser cooling of atoms with broadband real Gaussian laser fields. Physical Review A 1992 May 1;45(9):6522-6538.

24. Atutov SN, Calabrese R, Grimm R, et al. ``White-light'' Laser Cooling of a Fast Stored Ion Beam. Physical Review Letters 1998 Mar 9;80(10):2129-2132.

25. An K, Yang C, Dasari RR and et al.Cavity ring-down technique and its application to the measurement of ultraslow velocities. Optics Letters 1995 May 1;20(9):1068-1070.

26. Poirson J, Bretenaker F, Vallet M, and Floch AL. Analytical and experimental study of ringing effects in a FabryCPerot cavity : Application to the measurement of high finesses. J. Opt. Soc. Am. B, 1997 Nov ;14(11):2811-2817.

27. Yang W, Joshi A, and Xiao M. Enhancement of the cavity ringdown effect based on electromagnetically induced transparency. Optics Letters 2004 Sep 15;29(18):2133-2135.

Chapter 2. Phase-dependent atomic coherence and interference in multi-level atomic systems

Jiepeng Zhang and Yifu Zhu
Florida International University

Abstract: Atomic coherence and interference manifested by electromagentically induced transparency (EIT) and coherent population trapping (CPT) plays an important role in the current studies of atom-photon interactions and has found numerous applications in optical physics. EIT is created in a three-level atomic system by a coupling field and results in destructive interference between two excitation paths of a weak probe laser interacting with the atomic medium. This leads to suppressed linear absorption and rapidly varying atomic dispersion for the probe laser near the atomic resonance, which provides the platform for a variety of applications such as nonlinear optics at low light levels, slow light manipulation, and quantum state engineering for photons and atoms.

Here we extend the simple three-level EIT system to more complicated and versatile configurations in a multi-level atomic system coupled by multiple laser fields. We show that with multiple excitation paths provided by different laser fields, phase-dependent quantum interference is induced: either constructive or destructive interference can be realized by varying the relative phases among the laser fields. Two specific examples are discussed. One is a three-level system coupled by bichromatic coupling and probe fields, in which the phase dependent interference between the resonant two-photon Raman transitions can be initiated and controlled. Another is a four-level system coupled by two coupling fields and two probe fields, in which a double-EIT configuration is created by the phase-dependent interference between three-photon and one-photon excitation processes. We analyze the coherently coupled multi-level atomic system and discuss the control parameters for the onset of constructive or destructive quantum interference. We describe two experiments performed with cold Rb atoms that can be approximately treated as the coherently coupled three-level and four-level atomic systems respectively. The experimental results show the phase-dependent quantum coherence and interference in the multi-level Rb atomic system, and agree with the theoretical calculations based on the coherently coupled three-level or four-level model system.

Introduction

Atomic coherence and interference has been a subject of intense theoretical and experimental studies in recent years due to its importance in the fundamental understanding of atom-photon interactions and practical applications in optical physics. A powerful technique that coherently prepares atomic systems and generates quantum coherence and interference is through electromagnetically induced transparency (EIT) and Coherent population trapping (CPT) [1-2]. The basic EIT system consists of a three-level Λ-type atomic system coupled by two monochromatic light fields as shown in Fig. 1(a); a strong coupling laser resonantly drives one atomic transition and creates destructive quantum interference for the other transition coupled by a weak probe laser. The destructive interference suppresses the resonant absorption of the weak probe laser and leads to a steep normal dispersion near the atomic resonance. A variety of applications utilizing the EIT atomic system have been explored recently, including quantum state manipulation, coherent spectroscopic measurements, nonlinear optics at low light levels, ultra-slow group velocity for light propagation, coherent light storage and recall, and quantum information processing [3-51].

EIT can be understood from the dressed-state picture: the strong coupling laser connecting the atomic states |2> and |3> in the three-level Λ-type system creates two dressed states that consist of the coherent superposition of the states |2> and |3> and is given by $|+\rangle = \frac{1}{\sqrt{2}}(|3\rangle + |2\rangle)$ and $|-\rangle = \frac{1}{\sqrt{2}}(|3\rangle - |2\rangle)$ (Fig. 1(b)). The two dressed states are separated in frequency (energy) by the Rabi frequency of the coupling laser and result in two transition paths (|1>→|+> and |1>→|->) for the probe laser (Fig. 1(b)). On the resonance frequency of the bare state transition |1> → |3>, the quantum interference between the two transition paths is destructive due to the opposite frequency detuning and leads to the suppression of the probe light absorption, rendering the three-level atomic medium transparent to the probe laser. Since the two transitions are linear and are connected by the same probe laser field, the quantum interference is independent of the probe laser phase and is destructive only.

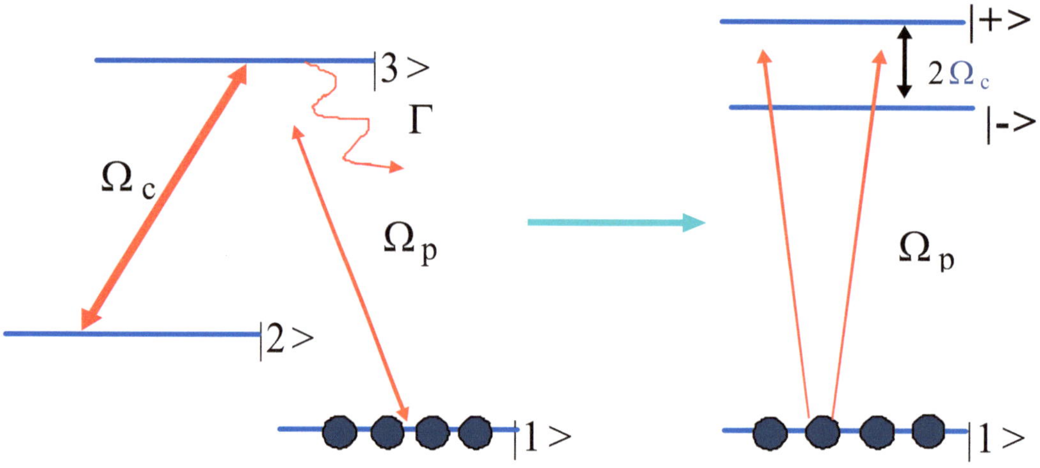

Fig.1 (a) Three-level Λ-type system coupled by two laser fields. (b) The dressed-state picture shows the two transition channels for the probe laser.

The atomic coherence and interference phenomenon in the simple three-level system such as EIT can be extended to more complicated multi-level atomic systems. A variety of other phenomena and applications involving three or four-level EIT systems have been studied in recent years. In particular, phase-dependent atomic coherence and interference has been explored [52-66]. These studies show that in multi-level atomic systems coupled by multiple laser fields, there are often various types of nonlinear optical transitions involving multiple laser fields and the quantum interference among these transition paths may exhibit complicated spectral and dynamic features that can be manipulated with the system parameters such as the laser field amplitudes and phases. Here we present two examples of such coherently coupled multi-level atomic systems in which the quantum interference is induced between two nonlinear transition paths and can be controlled by the relative phase of the laser fields.

The first example is a three-level Λ-type system coupled by bichromatic coupling and probe fields, which opens two Raman transition channels [60]. The phase dependent interference between the resonant two-photon Raman transitions depends on the relative phases of the laser fields; either constructive interference or destructive interference between the two Raman channels can be obtained by controlling the laser phases. The second example is a four-level system coupled by two coupling fields and two probe fields, in which a double-EIT configuration is created by the phase-dependent interference between the three-photon and one-photon excitation processes, or equivalently two independent Raman transition channels [58,62]. We will provide theoretical analyses of the phase dependent quantum interference in the two multi-level atomic systems and present experimental results obtained with cold Rb atoms. The two systems provide basic platforms to study coherent atom-photon interactions and quantum state manipulations, and to explore useful applications of the phase-dependent interference in the multi-level atomic systems.

Phase-dependent atomic coherence and interference in a single Λ-type system

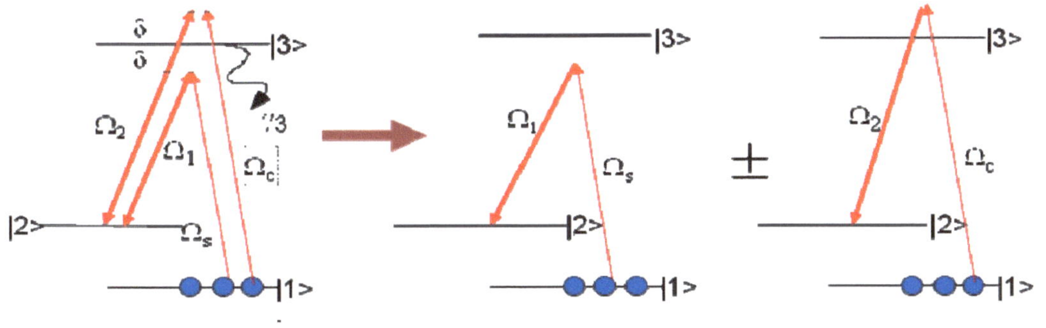

Fig.2 (a) Three-level Λ-type system coupled by four laser fields, two coupling field with Rabi frequencies Ω_1 and Ω_2 and two probe fields (signal and control) with Rabi frequencies Ω_c and Ω_s. The frequency separation the two coupling fields as well as the frequency separation of the two probe fields are matched symmetrically at the resonance. (b) Two resonant Raman channels are induced by the four laser fields and interfere with each other, leading to the enhanced or suppression Raman transitions.

Fig. 2(a) depicts a three-level Λ–type atomic system coupled by four laser fields. Two coupling fields with frequencies ν_1 and ν_2 (ν_2-ν_1≡2δ) drive the atomic transition |2>-|3> with Rabi frequencies Ω_1 and Ω_2 respectively. Two weak probe fields, one signal ν_s and one control field ν_c, couple the atomic transition |1>-|3> with Rabi frequency Ω_s and Ω_c respectively. The frequency difference is fixed by ν_c-ν_s=2δ. Here $\Omega_i =|\Omega_i| e^{i\phi_i}$ (i=c, s, 1, and 2) is characterized by the amplitude $|\Omega_i|$ and phase ϕ_i. The frequency detunings are chosen as $\Delta_1 = \nu_1 - \nu_{32} = -\delta$ and $\Delta_2 = \nu_2 - \nu_{32} = \delta$ for the two coupling fields. For the two probe fields, the frequency detunings are defined as $\Delta_c = \nu_c - \nu_{31} - \delta$ and $\Delta_s = \nu_s - \nu_{31} + \delta$. Under the condition of strong coupling fields and weak probe fields, i.e., $|\Omega_{c(s)}| << |\Omega_{1(2)}|$, the atomic population is concentrated in the state |1>. The two coupling fields and the two probe fields induce multiple transition channels for the two-photon Raman process |1>-|3>-|2>. We note that the same two-photon Raman process is responsible for the standard EIT in the three-level Λ system coupled by monochromatic coupling and probe fields, but only one Raman channel exists there. EIT/CPT is induced when the two-photon Raman transition is resonant and a dark state (consisting of coherent superposition of ground states |1> and |2>) is created [1-2]. When the same three-level Λ system is coupled by two bichromatic fields, multiple Raman transition channels, $|1> \xrightarrow{\nu_{c(s)}} |3> \xrightarrow{\nu_{1(2)}} |2>$, are opened and quantum interference among the multiple channels occur. The amplitudes of the Raman transition channels depend on the Raman transition moments and can be controlled by varying the relative phases among the laser fields. When the probe field detunings satisfy $\Delta_c = \nu_c - \nu_{31} - \delta = 0$ and $\Delta_s = \nu_s - \nu_{31} + \delta = 0$, there are two resonant Raman transition channels, $|1> \xrightarrow{\nu_s} |3> \xrightarrow{\nu_1} |2>$ and $|1> \xrightarrow{\nu_c} |3> \xrightarrow{\nu_2} |2>$, and two non-resonant Raman channels $|1> \xrightarrow{\nu_s} |3> \xrightarrow{\nu_2} |2>$ and $|1> \xrightarrow{\nu_c} |3> \xrightarrow{\nu_1} |2>$. If $\delta >> |\Omega_{1(2)}|$, the two non-resonant Raman channels can be neglected and the system is characterized by the two resonant Raman channels the amplitudes of which are proportional to $\frac{n(\rho_{11} - \rho_{22})\Omega_s \Omega_2^*}{(\nu_{31} - \nu_s)^2 (\nu_{21} - \nu_c + \nu_2 - i\gamma)}$ and $\frac{n(\rho_{11} - \rho_{22})\Omega_c \Omega_1^*}{(\nu_{31} - \nu_c)^2 (\nu_{21} - \nu_c + \nu_1 - i\gamma)}$ respectively [67].

For the resonant coupling depicted in Fig. 2(a) and assuming the two resonant Raman channels have the same amplitude, the combined two-photon Raman transition probability is proportional to $(1 + \cos(\phi_s - \phi_c + \phi_1 - \phi_2))$. One can vary the relative laser phase difference $\Delta\phi = \phi_s - \phi_c + \phi_1 - \phi_2$ to obtain either constructive interference or destructive interference between the open resonant Raman channels. When the two Raman channels are resonant with the two-photon transition |1>-|3>-|2>, the destructive interference among them will create dark states and suppresses the light absorption (CPT type transparency). On the other hand, if the interference is constructive, the two-photon Raman transition is enhanced and the probe light fields are attenuated. Under appropriate amplitude and phase conditions, the interference contrast can reach 100%, thus complete transparency for the destructive interference or maximum absorption for the constructive interference can be obtained in the coherently Λ system. We note that to observe the quantum interference discussed here, it only requires the phase locking between the two coupling fields and also the separate phase locking between the two probe fields. It is not necessary for the four laser fields to have a locked phase relationship.

For a quantitative analysis of the three-level Λ-type atomic system coupled by four laser fields, we start with the interaction Hamiltonian

$$H = \sum_{j=1}^{3} \varepsilon_j |j\rangle\langle j| - \left(\Omega_c e^{-i\theta_c} |3\rangle\langle 1| + h.c.\right) - \left(\Omega_s e^{-i\theta_s} |3\rangle\langle 1| + h.c.\right) - \left(\Omega_1 e^{-i\theta_1} |3\rangle\langle 2| + h.c.\right) - \left(\Omega_2 e^{-i\theta_2} |3\rangle\langle 2| + h.c.\right), \qquad (1)$$

The Hamiltonian describes a four-level system coupled by two bichromatic laser fields and the equations of motion can be derived with the density matrix method using Liouville equations $\dot{\rho} = \frac{1}{i\hbar}[H, \rho]$. It is difficult to derive a closed form analytical solution for the density matrix but the equation can be solved numerically by expanding the density matrix elements ρ_{ij} (i,j=1-3) in term of harmonic components of the coupling fields and the probe fields ($\rho_{ij}(t) = \sum_{n=-\infty}^{\infty} \rho_{ij}^{n} e^{in\delta}$) and truncating the order n at a sufficiently large value [64-66].

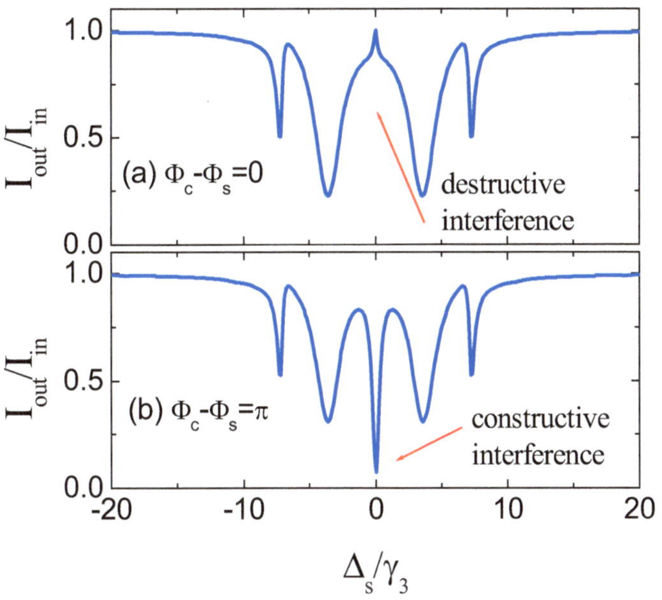

Fig. 3 calculated transmission spectrum of the signal field versus the signal frequency detuning Δ_s while the frequency detunings of the two coupling fields are kept at $\Delta_1 = \nu_1 - \nu_{32} = -\delta$ and $\Delta_2 = \nu_2 - \nu_{32} = \delta$. Other parameters used in the calculations are $\delta=4\gamma_3$, $\Omega_1=\Omega_2=\gamma_3$, $\Omega_s=\Omega_c=0.1\ \gamma_3$, and $\phi_1=\phi_2$.

Fig. 3 plots the calculated signal light transmission versus the frequency detuning Δ_s under the condition of the symmetric resonant excitations, i.e. $\Delta_1 = \nu_1 - \nu_{32} = -\delta$ and $\Delta_2 = \nu_2 - \nu_{32} = \delta$, and $\Omega_1=\Omega_2=\gamma_3$, $\Omega_s=\Omega_c=0.1\ \gamma_3$, and $\phi_1=\phi_2$. Under such conditions, the weak control field has the identical absorption spectrum as the signal field. Fig. 3 shows that at the two-photon Raman resonance, $\Delta_s = \nu_s - \nu_{31} + \delta = 0$ ($\Delta_c = \nu_c - \nu_{31} - \delta = 0$ as well), the signal field absorption is suppressed by the destructive interference between the two resonant Raman channels when the signal field and the control field have the same phase (ϕ_s-ϕ_c=0); but when the signal field and the control field are out of phase (ϕ_s-ϕ_c=π), the signal field absorption is enhanced by the constructive interference between the two resonant Raman channels.

There are additional spectral features at $\Delta_s \neq 0$ associated with the three-level Λ-type system which can be attributed to the ladder-type dressed states created by the two coupling field. As matter of fact, the peak structure of the signal absorption spectrum represents the ladder type structure of the dressed states created by a strong bichromatic field [68-72]. The dressed states are separated equally by the frequency interval of 2δ and the oscillation strength of the signal transition |1>-|3> is distributed among the dressed state manifold: the stronger the bichromatic coupling, the more peaks are observed in the signal spectrum representing more dressed states excitation in the dressed state manifold. However, since the oscillation strength is a constant for a given atomic transition |1>-|3> and if it is distributed in more dresses states, the absorption peak amplitudes will decrease as the bichromatic coupling field becomes stronger [68-71].

We observed the phase-dependent interference of the three-level Λ-type atomic system coupled by the multiple laser fields. The experiment is done with cold ^{85}Rb atoms confined in a magneto-optical trap (MOT). The MOT is obtained with a tapered-amplifier diode laser (Toptica TA-100) used as the cooling and trapping laser, and an extended-cavity diode laser (output power ~30 mW) is used as the repump laser. The trapped ^{85}Rb atom cloud is ~ 2 mm in diameter and the measured optical depth $n\sigma_{13}\ell$ is ~ 6. Fig. 4(a) depicts the laser coupling scheme of the three-level Λ-type system realized with the ^{85}Rb D1 transitions and Fig. 4(b) shows the simplified diagram of the experimental set up. An extended-cavity diode laser with a beam diameter ~ 3 mm and output power ~ 50 mW is used as the coupling laser. The driving electric current to the diode laser is modulated at 2δ=80 MHz, which produces two first-order frequency sidebands (ν_+ and ν_-) with about equal amplitudes ($|\Omega_\pm| \sim |\Omega_0| \sim 6$ MHz) as the central carrier (ν_0). The carrier ν_0 and of the sideband (ν_+ or ν_-) are used as the two coupling fields ν_1 and ν_2 respectively. As the higher-order sidebands are weaker than the carrier and are detuned farther away from the atomic resonance, their effects on the coupled Λ system can be neglected under our experimental conditions. Another extended-cavity diode laser with a beam diameter ~ 0.5 mm passes through an acousto-optics modulator (AOM2). The first-order diffracted beam (shifted in frequency by 2δ=80 MHz) is used as the control field and the zeroth-order beam is used as the signal (probe) field. The zeroth beam is passed through an electro-optic modulator (EOM, New Focus 4002) and its phase is controlled by a DC voltage applied to the EOM. The two probe beams with about equal powers are combined in a beam combiner and then are coupled into a polarization-maintaining single-mode fiber, the output of which are collimated, attenuated, and then directed to be overlapped with the frequency-modulated coupling laser in the MOT (with the propagating directions separated by a small angle of ~2°). After transmission through the MOT, the bichromatic probe light (both the signal field and the control field) is collected by a photodiode detector. All laser fields are circularly polarized (σ^+) and interact with the Rb

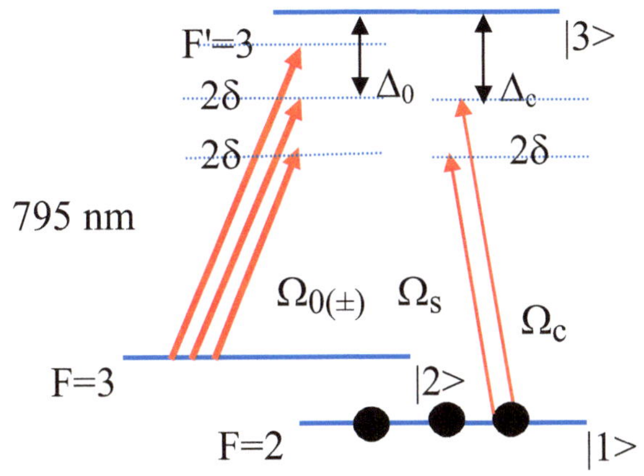

Fig4(a)

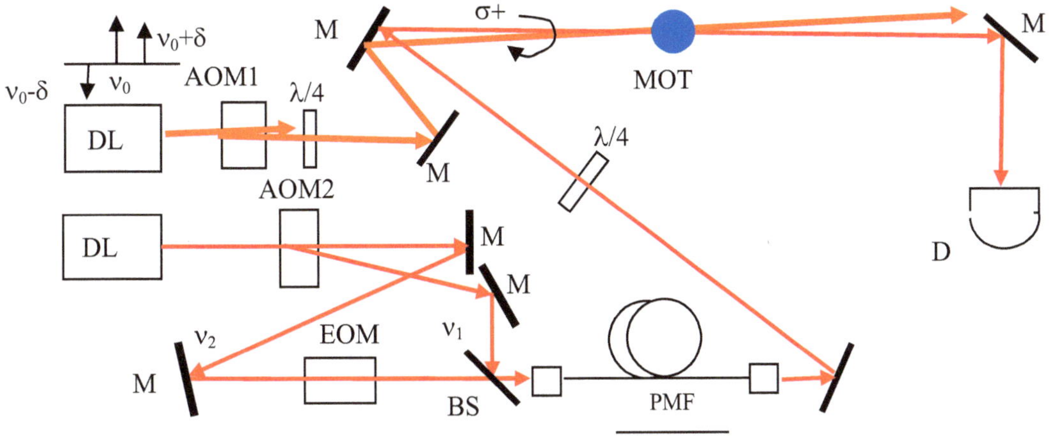

Fig4(b)

Fig. 4 (a) The ^{85}Rb D1 transitions coupled by a frequency modulated coupling field and two probe fields (one as the signal Ω_s and another as the control Ω_c) behaves as the coherently coupled three-level Λ-type system. The spontaneous decay rate of the excited state |3> is γ_3 (=2πx5.4x10^6 s^{-1} for the $5P_{1/2}$ F'=3 state). (b) Simplified diagram of the experimental set up. AOM: acousto-optic modulator; EOM: electro-optic modulator; PMF: polarization maintaining fiber; λ/4: quarter-wave plate; DL: extended-cavity diode laser; M: mirror; D: photodetector.

transitions to form 4 separate sets of the three-level Λ system among the magnetic sublevels.

The experiment is run in a sequential mode with a repetition rate of 5 Hz. all lasers are turned on or off by Acousto-Optic Modulators (AOM) according to the time sequence described below. For each period of 200 ms, ~198 ms is used for cooling and trapping of the ^{85}Rb atoms, during which the trapping laser and the repump laser are turned on by two AOMs while the coupling laser and the weak laser are off. The time for the data collection lasts ~ 2 ms, during which the trapping laser and the repump laser are turned off as well as the current to the anti-Helmholtz coils of the MOT, and the coupling laser and the weak laser are turned on. For the spectral measurements, the weak laser frequency is scanned across the ^{85}Rb D_1 F=2→F' transitions after a 0.1 ms delay and the transmission of the weak laser and the fluorescent photons (proportional to the excited-state population) are then recorded.

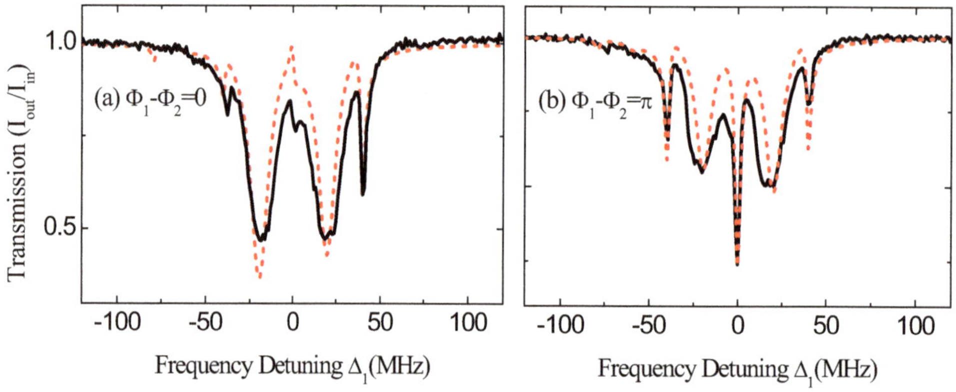

Fig. 5. Measured intensity transmission of the signal and control light versus the signal detuning Δ_s while the carrier component of the coupling laser is detuned by Δ_0=-20 MHz (δ=40 MHz). (a) ϕ_1-ϕ_2=0 the destructive interference leads to the suppressed light absorption at Δ_S=0. (b) ϕ_1-ϕ_2=π, the constructive interference leads to the enhanced light absorption at Δ_S=0. Solid (dashed) lines are experimental data (calculations). The experimental parameters are $\Omega_1/(2\pi)\approx\Omega_2/(2\pi)\approx 6.5$ MHz, $\Omega_p/(2\pi)\approx\Omega_c/(2\pi)\approx 0.1$ MHz, which are used in the numerical calculations.

EIT in a Λ system is manifested by the dark state induced by the resonant two-photon Raman coupling of the ground states |1> and |2>. With a monochromatic coupling and probe fields, only one Raman channel exists and the dark state is always formed, which results in suppression of the probe absorption. In the Λ system coupled by the frequency-modulated coupling field and the bichromatic probe field, there are multiple Raman channels. Depending on the relative phases among the laser fields and the frequency detunings, either destructive interference or constructive interference can be obtained, which leads to either the dark state (suppressed light absorption) or the bright state (enhanced light absorption). Two values of the frequency detuning for the carrier component of the frequency modulated coupling field fulfill the requirement for the realization of the coupled three-level Λ-type system depicted in Fig. 2. First, at the carrier detuning Δ_0=δ, the carrier component of the frequency modulated coupling field plays the role of Ω_2 and the lower side band plays the role of Ω_1. The blue sideband ν_+ of the frequency modulated coupling field is further detuned from the atomic resonance by 3δ and does not form the resonant Raman channel. To a good approximation, only two resonant Raman channels, ($|1\rangle \xrightarrow{\nu_s} |3\rangle \xrightarrow{\nu_-} |2\rangle$ and $|1\rangle \xrightarrow{\nu_c} |3\rangle \xrightarrow{\nu_0} |2\rangle$), need to be considered; second, at the carrier detuning Δ_0=-δ, the carrier component of the frequency modulated coupling plays the role of Ω_1 and the blue sideband plays the role of Ω_2. The red sideband ν_- of the frequency modulated coupling field is further detuned from the atomic resonance by -3δ and does not form the resonant Raman channel. Then only two resonant Raman channels, ($|1\rangle \xrightarrow{\nu_s} |3\rangle \xrightarrow{\nu_0} |2\rangle$ and $|1\rangle \xrightarrow{\nu_c} |3\rangle \xrightarrow{\nu_+} |2\rangle$), need to be considered. Fig. 5 plots the probe transmission versus the signal detuning Δ_s for the fixed carrier detuning $\Delta_0=-\delta$. When the phase difference $\phi_c-\phi_s=0$ (Fig. 5a), the two Raman channels interfere destructively, and the EIT type spectrum is observed, exhibiting a transparency widow at $\Delta_s\approx 0$. When the phase difference $\phi_1-\phi_2=\pi$ (Fig. 5b), the two Raman channels interfere constructively, and a large absorption peak (~70%) is observed at $\Delta_s\approx 0$. The observed linewidth of the absorption peak at $\Delta_s\approx 0$ is ~ 4 MHz and is smaller than the natural linewidth of the Rb D1 transitions (~5.4 MHz). The central peak linewidth is smaller than that of the sideband absorption peaks at $\Delta_1\approx\pm\delta$ (determined by the decay rate of the excited state |3>) and is a characteristic of the atomic interference. Our calculations indicate that the observed linewidth is limited by the power broadening of the coupling field. With a much weaker coupling field, the linewidth of the enhanced absorption peak at $\Delta_s=0$ is determined by the decoherence rate of the ground state coherence and can be orders of magnitude smaller than the natural linewidth of the excited state |3>. The small asymmetry in the observed spectra of Fig. 4 can be attributed to the detuned sideband ν_+, which is confirmed by theoretical calculations replacing the three components coupling field with a bichromatic coupling field.

Phase-dependent atomic coherence and interference in a double Λ-type system

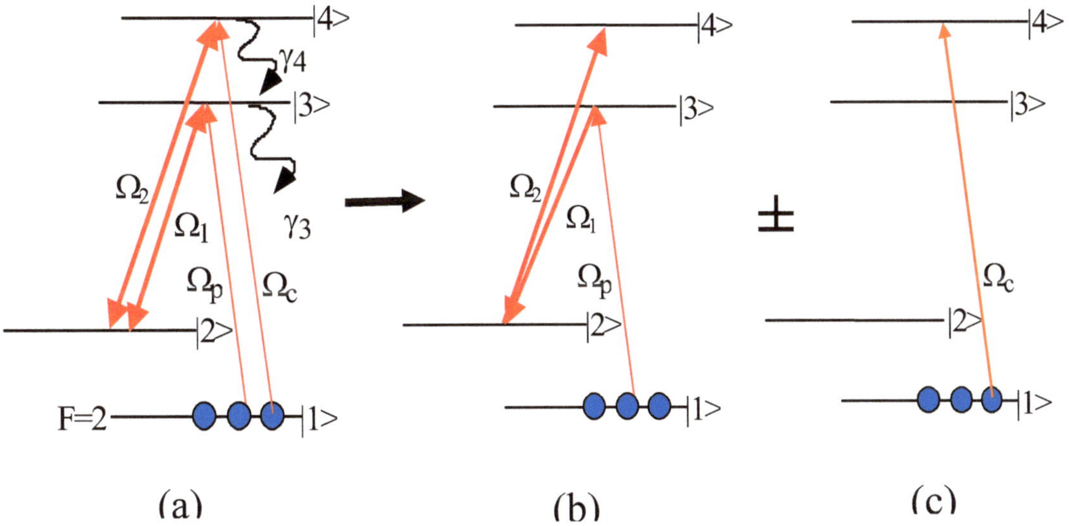

Fig. 6 (a) Four-level Double Λ-type atomic system coupled by four laser fields. γ_3 (γ_4) is the spontaneous decay rate ($\gamma_3\approx\gamma_4$ $=2\pi\text{x}5.4\text{x}10^6$ s^{-1}). Interference occurs between (b) three-photon excitation |1>-|3>-|2>-|4> and (c) one-photon excitation |1>-|4>. The interference between three-photon excitation |1>-|4>-|2>-|3> and one-photon excitation |1>-|3> is not drawn here.

The phase independent quantum interference through the multiple Raman transitions in the three-level Λ-type system can also be realized in a four-level double Λ-type system with a similar laser coupling scheme depicted in Fig. 6. The four-level double-Λ-type system is driven by two coupling fields. The coupling field 1 (2) drives the transition |2>-|3> (|2>-|4>) with Rabi frequency Ω_1 (Ω_2). Two weak laser fields, one as signal (probe) and another as control, drive the transitions |1>-|3> and |1>-|4> with Rabi frequency Ω_s and Ω_c respectively. The signal and control are interchangeable. Here $\Omega_i = |\Omega_i| e^{i\phi_i}$ (i=1,2, s, and c) is characterized by the amplitude $|\Omega_i|$ and phase ϕ_i. The frequency detunings for the respective transitions are defined as $\Delta_s = \omega_s - \omega_{31}$, $\Delta_c = \omega_c - \omega_{41}$, $\Delta_1 = \omega_1 - \omega_{32}$, and $\Delta_2 = \omega_2 - \omega_{42}$ (ω_i (i=s, 1, c, 2) is the angular frequency of the laser field i). The Hamiltonian for the coupled four-level double Λ system is given by

$$H = \sum_{j=1}^{4} \varepsilon_j |j\rangle\langle j| - \left(\Omega_c e^{-i\theta_c} |4\rangle\langle 1| + h.c.\right) - \left(\Omega_s e^{-i\theta_s} |3\rangle\langle 1| + h.c.\right)$$
$$- \left(\Omega_1 e^{-i\theta_1} |3\rangle\langle 2| + h.c.\right) - \left(\Omega_2 e^{-i\theta_2} |4\rangle\langle 2| + h.c.\right), \quad (2)$$

From the interaction Hamiltonian H, one can derive Schrödinger equation in the interaction picture, $i\frac{\partial}{\partial t}|\Psi\rangle = H|\Psi\rangle$. The susceptibilities at the signal frequency and the control frequency can be obtained by solving the Schrodinger equations. The Maxwell equations for the propagating weak signal field and the weak control field in the four-level double Λ-type medium of length ℓ can then be derived [62]. For $\Omega_1 \sim \Omega_2$, $\Omega_s \sim \Omega_c$, $\Omega_{1(2)} >> \Omega_{c(s)}$, $\Delta_c = \Delta_s$, and the four laser fields propagate in the z direction (neglecting depletion of the two coupling fields), the Maxwell equations for the signal field and the control field are given by

$$\frac{d\Omega_s}{dz} = \frac{iK_{13}(|\Omega_1|^2 - \delta_1\delta_c)}{\Lambda}\Omega_s - \frac{iK_{13}\Omega_1\Omega_2^*}{\Lambda}\Omega_c \quad (3a)$$

$$\frac{d\Omega_c}{dz} = \frac{iK_{14}(|\Omega_2|^2 - \delta_1\delta_s)}{\Lambda}\Omega_c - \frac{iK_{14}\Omega_1^*\Omega_2}{\Lambda}\Omega_s. \quad (3b)$$

Here $\delta_s = \Delta_s + i\gamma_3$, $\delta_1 = \Delta_s - \Delta_1 + i\gamma_2$, $\delta_c = \Delta_c + i\gamma_4$, and $\Lambda = \delta_1\delta_2\delta_s - \delta_s|\Omega_2|^2 - \delta_2|\Omega_1|^2$, and $K_{ij} = \frac{2\pi N\omega_{ij}|\mu_{ij}|^2}{\hbar c}$ (N is the atomic density). Eq. (3a) and (3b) can be solved analytically. Fig. 7 plots the calculated signal intensity transmission versus the frequency detuning Δ_s in the four-level system. Without the control field (Fig. 7(a)), the probe field is attenuated in the medium with peak absorptions at $\Delta_s = \pm\Omega = \pm\sqrt{\Omega_1^2 + \Omega_2^2}$ and Δ_s=0 (EIT enhanced nonlinear absorption [6]). When the control field is present and $\Omega_s(0)\Omega_2 = \Omega_c(0)\Omega_1$ ($\Omega_{s(c)}(0)$ is the incident signal (control) Rabi frequency at z=0), the absorption for the signal and the control at $\Delta_s=\Delta_c=0$ are suppressed (destructive interference) while the absorption for the signal and the control at $\Delta_s=\Delta_c\approx\pm\Omega$ are enhanced (constructive

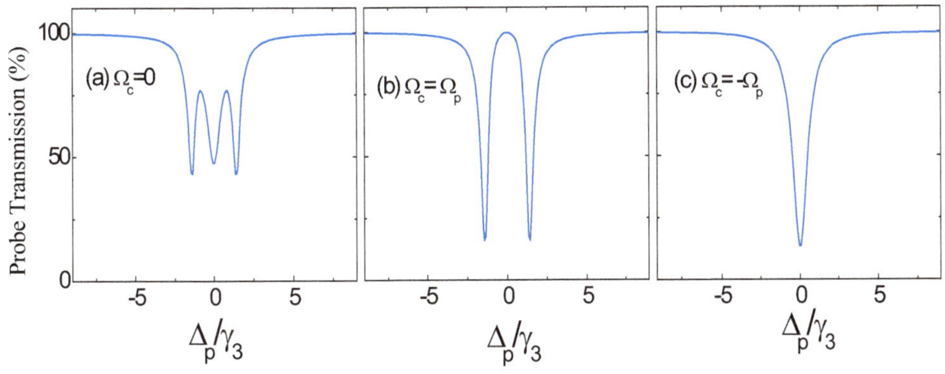

Fig. 7. (a) Calculated signal light transmission versus Δ_s without the control laser. (b) Calculated signal transmission versus $\Delta_s=\Delta_c$ with $\Omega_c(0)=\Omega_p(0)$. The absorption is suppressed by destructive interference at $\Delta_s=0$ and enhanced by constructive interference at $\Delta_s=\pm\Omega$. (c) Calculated probe absorption versus $\Delta_s=\Delta_c$ with $\Omega_c(0)=-\Omega_p(0)$. The absorption is enhanced by constructive interference at $\Delta_s=0$ and suppressed by destructive interference at $\Delta_s=\pm\Omega$. The parameters are $\Omega_1=\Omega_2=\gamma_3$, $\gamma_3=\gamma_4$, $\gamma_2=0.02\gamma_3$, $\Delta_1=\Delta_2=0$, and $K_{13}\ell/\gamma_3=K_{14}\ell/\gamma_4=1$.

interference) (Fig. 7b). That is, EIT is induced simultaneously for both the probe and control fields at $\Delta_s=\Delta_c=0$. For simplicity, we choose $\Omega_1=\Omega_2$ so the signal field and the control field have the identical absorption profile. On the other hand, when the four laser fields satisfy the condition $\Omega_s(0)\Omega_2=-\Omega_c(0)\Omega_1$, the signal and control absorptions at $\Delta_s=\Delta_c=0$ are enhanced (constructive interference) while the probe and control absorptions at $\Delta_s=\Delta_c\approx\pm\Omega$ are suppressed (destructive interference) (Fig. 7(c)). That is, the interference can be changed from constructive into destructive by controlling the relative phase among the four laser fields. The interference also requires an amplitude matching condition, in which the control Rabi frequency satisfying $|\Omega_c(0)|=|\Omega_s(0)\Omega_2/\Omega_1|=\alpha|\Omega_s(0)|$ (α can be ≥ 1 or <1 depending on the ratio of the Rabi frequencies for the two coupling fields).

The dressed state picture provides a simple explanation for the phase dependent quantum interference of the four-level system coupled by the multiple laser fields. The two resonant coupling fields create a manifold of three dressed states, the semi-classical representation of which is given by

$$|+>=\frac{1}{\sqrt{2}}\{|1>-\frac{\Omega_1}{\Omega}|3>-\frac{\Omega_2}{\Omega}|4>\}, \tag{4a}$$

$$|0>=\frac{\Omega_2}{\Omega}|3>-\frac{\Omega_1}{\Omega}|4> \quad , \tag{4b}$$

$$|->=\frac{1}{\sqrt{2}}\{|1>+\frac{\Omega_1}{\Omega}|3>+\frac{\Omega_2}{\Omega}|4>\}, \quad (4c)$$

where $\Omega=\sqrt{\Omega_1^2+\Omega_2^2}$. The corresponding energy shifts (relative to the transition energy of the bare atomic states) are Ω, 0, and –Ω, respectively. With two weak fields Ω_c and Ω_s, the transition probability from the state |1> to the dressed states |+> and |-> is

$$P_{1\pm}\propto|\Omega_s\Omega_1+\Omega_c\Omega_2|^2=|\Omega_s|^2|\Omega_1|^2+|\Omega_c|^2|\Omega_2|^2 \\ +2|\Omega_s||\Omega_1||\Omega_c||\Omega_2|\cos(\Phi_2+\Phi_c-\Phi_1-\Phi_s),$$

and the transition probability from the state |1> to the dressed state |0> is

$$P_{10}\propto|\Omega_s\Omega_2-\Omega_c\Omega_1|^2=|\Omega_s|^2|\Omega_2|^2+|\Omega_c|^2|\Omega_1|^2 \\ -2|\Omega_s||\Omega_1||\Omega_c||\Omega_2|\cos(\Phi_2+\Phi_c-\Phi_1-\Phi_s).$$

Φ_i (i=1,2.c, and s) is the phase of the laser field. The interference between the two excitation paths can be manipulated by varying the amplitudes and phases of the four laser fields. For example, when$\Delta\phi=\phi_s-\phi_c+\phi_1-\phi_2=0$, (with Ω_1=-Ω_2 and Ω_c=-Ω_s, ($\Phi_1=\Phi_2+\pi$ and ($\Phi_s=\Phi_c+\pi$) for the frequency modulation scheme used in our experiment, see discussion below), complete destructive interference occurs for the transitions |1>- |0> while complete constructive interference occurs for the transitions |1>-|±> (Fig. 7b). When$\Delta\phi=\phi_s-\phi_c+\phi_1-\phi_2=\pi$, complete constructive interference occurs for the transitions |1>-|0> while complete destructive interference occurs for the transitions |1>-|±> (Fig. 7c).

The phase dependent quantum interference can be also revealed by looking at the population probabilities in the excited atomic states |3> and |4> under the conditions of $|\Omega_1|\sim|\Omega_2|$, $|\Omega_s|\sim|\Omega_c|$, $|\Omega_{1(2)}|>>|\Omega_{c(s)}|$, and $\Delta_i=0$ (i=s,1,c, and 2). Under such symmetric resonant excitations, the adiabatic excited-state populations are

$$P_3=\left|\frac{\Omega_s(|\Omega_2|^2+\gamma_2\gamma_4)-\Omega_1\Omega_2^*\Omega_c}{\gamma_2\gamma_3\gamma_4+\gamma_3|\Omega_2|^2+\gamma_4|\Omega_1|^2}\right|^2, \quad (5a)$$

$$P_4=\left|\frac{\Omega_c(|\Omega_1|^2+\gamma_2\gamma_3)-\Omega_2\Omega_1^*\Omega_s}{\gamma_2\gamma_3\gamma_4+\gamma_3|\Omega_2|^2+\gamma_4|\Omega_1|^2}\right|^2, \quad (5b)$$

where γ_2 is the decay rate of the ground-state coherence ρ_{12}. The first term in the nominator of Eq. (5a) ((5b)) represents the one-photon excitation |1>-|3> (|1>-|4>) while the second term represents the three-photon excitation |1>-|4>-|2>-|3> (|1>-|3>-|2>-|4>). The two excitation paths interfere with each other, which can be manipulated by varying the phases and amplitudes of the laser fields. When $\Omega_1\Omega_c=\Omega_2\Omega_s$ (neglecting γ_2, which is justified due to $|\Omega_{1(2)}|^2>>\gamma_2\gamma_{3(4)}$), the interference is destructive: P_3 and P_4 vanish, and the probe and control fields propagate in the medium without attenuation, EIT occurs in the double Λ-type system. When $\Omega_1\Omega_c=-\Omega_2\Omega_s$ (with a π phase change from the previous case), the interference is constructive:

P_3 and P_4 are maximized, and the probe and control fields are attenuated in the medium.

The phase dependent quantum interference in the coupled double Λ system can also be understood by considering the resonant two-photon Raman transitions from |1>-|2>. There are two resonant Raman channels; one Raman channel is | $|1> \xrightarrow{\nu_s} |3> \xrightarrow{\nu_1} |2>$ and another one is $|1> \xrightarrow{\nu_c} |4> \xrightarrow{\nu_2} |2>$. Unlike the Raman transition channels in the three-level Λ-type system of Fig.2, the two resonant Raman channels are connected through the resonant intermediate excited states by the one-photon transitions |1>-|3> and |1>-|4. The distinction between the two cases corresponds to the technical difference between EIT and CPT, in which EIT is created by the two-photon Raman resonance through the resonant intermediate excited state while CPT is created by the two-photon Raman resonance through the off-resonant intermediate excited state.

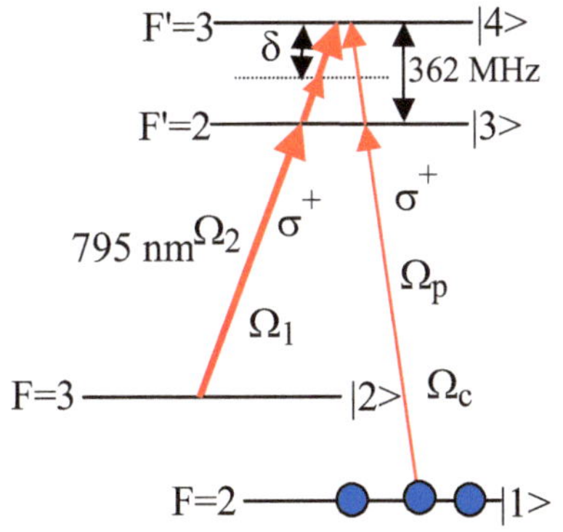

Fig. 8. The four-level double Λ-type system realized with the ^{85}Rb D1 transitions.

We observed the phase-dependent quantum interference in the double Λ system realized with cold ^{85}Rb atoms coupled by four laser fields. The coherently coupled four-level double Λ-type system realized with the laser coupling scheme for the ^{85}Rb D_1 transitions is shown in Fig. 8 and the simplified experimental set up is depicted in Fig. 4(b). An extended-cavity diode laser with a beam diameter ~ 3 mm and output power ~ 50 mW is used as the coupling laser. The driving electric current to the diode laser is modulated at δ=181 MHz with a modulation index ~0.5, which produces two first-order frequency sidebands separated by 362 MHz. The two sidebands are tuned to the ^{85}Rb D_1 F=3→F'=2 and F=3→F'=3 transitions respectively and serve as the two coupling fields ($\Omega_1 \approx -\Omega_2$ due to a π phase difference between the two sidebands). Another extended-cavity diode laser with a beam diameter ~ 0.5 mm passes through an acousto-optics modulator (AOM2). The first-order diffracted beam (shifted in frequency by δ=362 MHz) is used as the signal light and the zeroth-order beam is used as the control light, which are tuned to the D_1 F=2→F'=2 and F=2→F'=3 transitions, respectively. The control beam is passed through an electro-optic modulator (EOM, New Focus 4002) and its phase is varied by a DC voltage applied to the EOM. The two beams with about equal powers are combined in a beam combiner and then are coupled into a polarization-maintaining single-mode fiber, the output of which are collimated and passed through another AOM (AOM3). The first order diffracted beam containing both the signal light and the control light is then attenuated and directed to be overlapped with the frequency-modulated coupling laser in the MOT (with the propagating directions separated by a small angle of ~2°). The transmitted beam of the weak laser (both the signal and the control) is collected by a photodetector.

The intensity transmission of the weak probe laser versus the frequency detuning (Δ_c=Δ_s) are plotted in Fig. 9. Fig. 9 (a) presents the measurement without the control field and represents the EIT manifested absorption spectra in the four-level system observed before [39]. The light transmission spectrum with both the signal and control fields present is plotted in Fig. 9(b), which shows that the absorption at the resonance (Δ_p=Δ_c=0) is suppressed by the destructive interference, which results in simultaneous EIT for both the signal field and the control field.

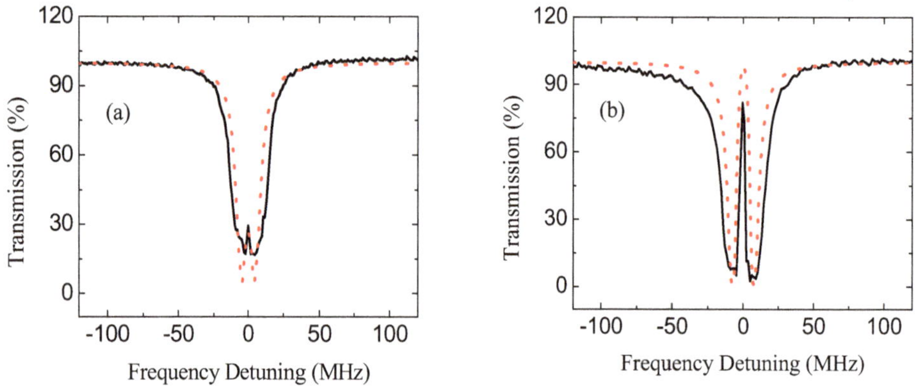

Fig. 9 Measured (solid lines) and calculated (dashed lines) transmission spectra of the combined signal and control fields versus the frequency detuning $\Delta_c = \Delta_s$. (a) No control laser is present. (b) Control laser is present and EIT is created in the double Λ system. The experimental parameters are $|\Omega_1|/(2\pi)\approx|\Omega_2|/(2\pi)\approx5.4$ MHz, $|\Omega_s|/(2\pi)\approx|\Omega_c|/(2\pi)\approx0.1$ MHz and $\Delta_1 = \Delta_2 = 0$. The fitting parameters are γ_2=0.02γ_3 and the optical depth $n\sigma_{13}\ell = 5$.

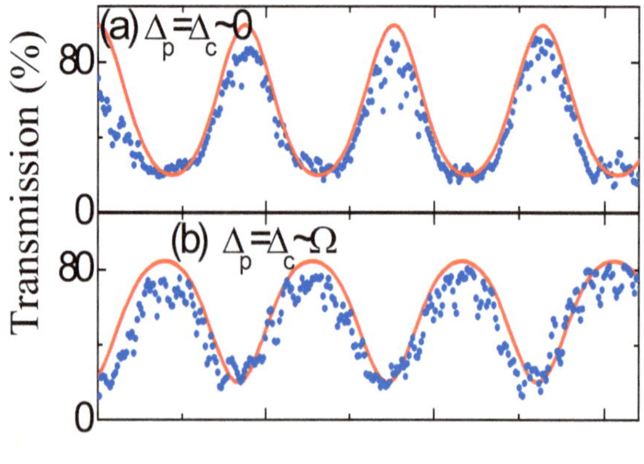

Fig. 10. Transmission of the control and the signal fields versus the phase variation of the control field. Dots are experimental data and solid lines are calculations with the adiabatic phase change. The control

laser phase is varied as $\Phi_c = \pi\cos(2\pi ft) + \varphi_0$ (f=2.3 KHz and φ_0 is set by the DC off-set voltage); (a) The light transmission at $\Delta_c = \Delta_s \approx 0$; (b) The light transmission at $\Delta_c = \Delta_s \approx \Omega$; $\Omega_1/(2\pi) \approx \Omega_2/(2\pi) \approx 4.5$ MHz, and the other parameters are the same as that in Fig. 9.

The phase-dependent interference is revealed by measuring the weak laser transmission versus the control laser phase. Fig. 10 ((a) and (b)) plots the transmitted signal and control beams versus the control laser phase Φ_c as Φ_c is varied by a sinusoidal voltage applied to the EOM. Fig. 10(a) plots the light transmission versus Φ_c at $\Delta_p = \Delta_c \approx 0$ and Fig. 10(b) plots the light transmission versus Φ_c at $\Delta_p = \Delta_c \approx \Omega$. The data show that there is a π phase difference in the interference pattern between the two cases, illustrating the phase and frequency control of the light transmission by the quantum interference.

Conclusion

Phase-dependent coherence and interference can be induced in a multi-level atomic system coupled by multiple laser fields. Two simple examples are presented here, a three-level Λ-type system coupled by four laser fields and a four-level double Λ-type system coupled also by four laser fields. The four laser fields induce the coherent nonlinear optical processes and open multiple transitions channels. The quantum interference among the multiple channels depends on the relative phase difference of the laser fields. Simple experiments show that constructive or destructive interference associated with multiple two-photon Raman channels in the two coherently coupled systems can be controlled by the relative phase of the laser fields. Rich spectral features exhibiting multiple transparency windows and absorption peaks are observed. The multicolor EIT-type system may be useful for a variety of application in coherent nonlinear optics and quantum optics such as manipulation of group velocities of multicolor, multiple light pulses, for optical switching at ultra-low light intensities, for precision spectroscopic measurements, and for phase control of the quantum state manipulation and quantum memory.

This paper is based upon work supported by the National Science Foundation under Grant No. 0757984. Jiepeng Zhang's current address is Physics Division P-23, Los Alamos National Laboratory, Los Alamos, NM 87544.

References

1. S. E. Harris, Phys. Today **50**, 36-42 (1997), and references therein.
2. E. Arimondo, in Progress in Optics, edited by E. Wolf (Elsevier Science, Amsterdam, 1996), p. 257-354, and references therein.
3. K. J. Boller, A. Imamoglu, and S. E. Harris, Phys. Rev. Lett. **66**, 2593 (1991).
4. J. Gao, C. Guo, X. Guo, G. Jin, P. Wang, J. Zhao, H. Zhang, Y. Jiang, D. Wang, and D. Jiang, Opt. Commun. **93**, 323(1992).
5. Y. Q. Li and M. Xiao, Phys. Rev. A **51**, R2703 (1995).
6. K. Hakuta, L. Marmet, and B. P. Soicheff, Phys. Rev. Lett. **66**, 596 (1991).
7. M. O. Scully, Phys. Rev. Letts. **66**, 1855 (1991).
8. S. Y. Zhu and M. O. Scully, Phys. Rev. Lett. **76**, 388(1996).

9. M. Xiao, Y. Li, S. Jin, and J. Gea-Banacloche, Phys. Rev. Lett. **74**, 666 (1995).
10. G. S. Agarwal and W. Harshaawardhan, Phys. Rev. Lett. **77**, 1039 (1996).
11. Harshawardhan W, Agarwal GS, Phys. Rev. A **58**, 598 (1998).
12. O. Schmidt, R. Wynands, Z. Hussein, and D. Meschede, Phys. Rev. A **53**, R27 (1996).
13. M. Jain, A. J. Merriam, K. Kasapi, G. Y. Yin, and S. E. Harris, Phys. Rev. Lett. **75**, 4385 (1995).
14. Y. Li and M. Xiao, Opt. Lett. **21**, 1064 (1996).
15. H. Lee, P. Polynkin, M. O. Scully, and S. Y. Zhu, Phys. Rev. A 4454(1997).
16. P. R. Hemmer, D. P. Kats, J. Donoghue, M. Cronin-Golomb, M. S. Shahriar, and P. Kumar, Opt. Lett. **20**, 982 (1995).
17. K. Hakuta, M. Suzuki, M. Katsuragawa, and J. Li, Phys. Rev. Lett, 79, 209(1997).
18. M. D. Lukin, P. R. Hemmer, M. Loffler, M. O. Scully, Phys. Rev. Lett **81**, 2675 (1998).
19. M. D. Lukin, Fleischhauer, M. O. Scully, Phys. Rev. Lett. **79**, 2959-2962 (1997).
20. M. O. Scully and M. Fleischhauer, Phys. Rev. Lett. **69**, 1360 (1992).
21. B. S. Ham, M. S. Shahriar, P. R. Hemmer, Opt. Lett. 24, 86 (1999).
22. D. Budker, D. F. Kimball, S. M. Rochester, V. V. Yashchuk, M. Zolotorev, Phys. Rev. A **62**, 3403(2000).
23. G. Morigi, J. Eschner, and C. H. Keitel, Phys. Rev. Lett. **85**, 4458 (2000).
24. C. F. Roos, D. Leibfried, A. Mundt, F. Schmidt-Kaler, J. Eschner, R. Blatt R, Phys. Rev. Lett. **85**, 5547 (2000).
25. L.V. Hau, S. E. Harris, and C. H. Behroozi, , Nature **397**, 594 (1999).
26. M. M. Kash, V. A. Sautenkov, A. S. Zibrov, L. Hollberg, G. R. Welch, M. D. Lukin, Y. Rostovtsev, E. S. Fry, and M. O. Scully, Phys. Rev. Lett. **82**, 5229 (1999).
27. D. Budker, D. F. Kimball, S. M. Rochester, and V. V. Yashchuk, Phys. Rev. Lett. **83**, 1767 (1999).
28. M. D. Lukin, P. R. Hemmer, and M. O. Scully, Adv. At. Mol. Opt. Phys. 42, 347(2000).
29. M. Fleischhauer and M. D. Lukin, Rev. Lett. **84**, 5094 (2000).
30. C. Liu, Z. Dutton, C. H. Behroozi, and L. Hau, Nature **409**, 490 (2001).
31. D. F. Phillips, M. Fleischhauer, A. Mair, R. L. Walsworth, and M. D. Lukin, Phys. Rev. Lett. **86**, 783 (2001).
32. A. S. Zibrov, A. B. Matsko, O. Kocharovskaya, Y. V. Rostovtsev, G. R. Welch, and M. O. Scully, Phys. Rev. Lett. **88**, 103601 (2002).
33. S. E. Harris and L. V. Hau, Phys. Rev. Lett. **82**, 4611 (1999).
34. M. D. Lukin and A. Imamoglu, Phys. Rev. Lett. 84, 1419 (2000).
35. M. D. Lukin and A. Imamoglu, Nature 413, 273 (2001).
36. B. S. Ham and P. R. Hemmer, Phy. Rev. Lett. 84, 4080(2000).
37. D. Vitali, M. Fortunato, and P. Tombesi, Phys. Rev. Lett. 85, 445 (2000).
38. D. Petrosyan and G. Kurizki, Phys. Rev. A 65, 033833(2002).
39. M. Yan, E. Ricky and Y. Zhu, Phys. Rev. A, 64, 041801(R)(2001).
40. A. Patnaik, J. Liang and K. Hakuta, Phys. Rev. A 66, 063808(2002).
41. H. Wang, D. Goorskey, and M. Xiao, Opt. Lett. **27**, 1354 (2002).
42. D. A. Braje, V. Balic, G. Y. Yin, and S. E. Harris, Phys. Rev. A **68**, 041801(R) (2003).
43. H. Schmidt and A. Imamoglu, Opt. Lett. 21, 1936(1996).
44. A. B. Matsko, I. Novikova, G. R. Welch, and M. S. Zubairy, Opt. Lett. **28**, 96 (2003).
45. H. Kang and Y. Zhu, Phys. Rev. Lett. 91, 93601(2003).

46. M. G. Payne and L. Deng, Phys. Rev. Lett. **91**, 123602 (2003).
47. T. Hong, Phys. Rev. Lett. 90, 183901 (2003).
48. Y. Wu and L. Deng, Phys. Rev. Lett. 93, 143904(2004).
49. H. Kang, G. Hernandez, and Y. Zhu, Phys. Rev. Lett. 93, 073601(2004).
50. Y. F. Chen, Z. H. Tsai, Y. C. Liu, and I. A. Yu, Opt. Lett. **30**, 3207(2005).
51. Y. F. Chen, C. Y. Wang, S. H. Wang, and I. A. Yu, Phys. Rev. Lett. **96**, 043603 (2006).
52. N. Georgiades, E. S. Polzik, and H. J. Kimble, Opt. Lett. 21, 1688(1996).
53. E. A. Korsunsky et al, Phys. Rev. A 59, 2302(1999).
54. J. H. Wu and J. Y. Gao, Phy. Rev. A **65**, 063807(2002).
55. S. Y. Gao, F. L. Li, and S. Y. Zhu, Phys. Rev. A **66**, 043806(2002).
56. A. F. Huss, et al, Phys. Rev. Lett. 93, 223601(2004).
57. X. Hu, G. Cheng, J. Zou, X. Li, and D. Du, Phys. Rev. A 72, 023803(2005)
58. L. Deng and M. G. Payne, Phys. Rev. A 71, 011803(R)(2005).
59. Q. He, B. Zhang, X. Wei, J. Wu, S. Kuang, and J. Gao, Phys. Rev. A 77, 063827(2008).
60. J. Zhang, J. Xu, G. Hernandez, X. Hu, and Y. Zhu , Phys. Rev. A **75**, 043810 (2007).
61. J. Zhang, G. Hernandez, and Y. Zhu, Optics Express **16**, 19113(2008).
62. H. Kang, G. Hernandez, J. Zhang, and Y. Zhu, Phys. Rev. A **73**, 011802(R) (2006).
63. J. Zhang, G. Hernandez, and Y. Zhu, Opt. Lett. **32**, 1317(2007).
64. X. M. Hu, J. H. Zou, X. Li, D. Du, and G. L. Cheng, J. Phys. B At. Mol. Opt. Phys. 38, 683(2005).
65. X. X. Li, X. M. Hu, W. X. Shi, Q. Xu, H. J. Guo, and J. Y. Li, Chin. Phy. Lett. 23, 340(2006).
66. P. Li, T. Nakajima, and X.Ning, Phy. Rev. A **74**, 043408 (2006).
67. Y. R. Shen, The Principles of nonlinear optics, John Wiley & Sons, New York, 1984.
68. Y. Zhu, Q. Wu, A. Lezama, D. J. Gauthier, and T. W. Mossberg, Phys. Rev. A **41**, 6574 (1990).
69. G. S. Agarwal, Y. Zhu, Daniel J. Gauthier, and T. W. Mossberg, J. Opt. Soc.Am. B **8**, 1163(1991).
70. Z Ficek and H. S. Freedhoff, Progress in Optics 40, 389 (2000).
71. Z. Ficek , J. Seke, A. V. Soldatov, G. Adam , N. N. Bogolubov, Opt. Commun. 217, 299(2003).
72. J. Wang, Y. Zhu, K. J. Jiang, and M. S. Zhan, Phys. Rev. A 68, 063810 (2003).

Chapter 3. Atomic Localization and its application in atom nano-lithograph

Luling Jin[1], Yueping Niu[1], Hui Sun[1,2], Shiqi Jin[1] and Shangqing Gong[1]
1)Shanghai Institute of Optics and Fine Mechanics, Chinese Academy of Sciences
2)National University of Singapore

Abstract: In this chapter, we give a brief review of our recent research work related to atomic localization via the effects of atomic coherence and quantum interferences, as well as its application in atom nano-lithograph, which includes atomic localization based on double-dark resonance effects, sub-half-wavelength localization via two standing-wave fields and an atom nano-lithograph scheme via two orthogonal standing-wave fields, etc..

Introduction

In recent years, the precision position measurement of an atom passing through a standing-wave field has attracted considerable attention [1–28], because the study of atom localization affords potential applications in trapping of neutral atoms, laser cooling [29], atom nano-lithography [30], etc. The dynamics of atomic systems are different at different positions in the standing-wave field, because the strength of the interaction in a standing-wave field is position dependent. Thus the measurement of position -dependent quantities can provide information on the atomic position and lead to atom localization.

It is well known that quantum coherence and interference can give rise to some interesting phenomena, such as electromagnetically induced transparency (EIT) [31–33], slow light [34], giant Kerr nonlinearity [35, 36], four wave mixing [37], ultraslow optical solitons [38], and so on. Many atom localization schemes based on atomic coherence and quantum interference has been proposed. Some schemes of atom localization, based on the fact that the strength of the atom-field interaction depends on the position of the atom in the field, have been proposed using the phase shift of either the standing wave or the atomic dipole [1–3], the entanglement between the atom's position and its internal state [4], or other methods [5, 6]. Paspalakis and Knight [18, 19] put forward a scheme for localizing a three-level Λ-type atom in a standing-wave field. They showed that the measurement of upper state populations can lead to subwavelength localization, and simplified the demands on the initial-state preparation of an atom form the schemes of Zubairy and co-workers [8–10]. Agarwal and Kapale realized atom localization in a Lambda configuration [20]. In their scheme, the atom can be localized just at the nodes of the standing-wave field, which can be employed to obtain new kind of optical lattices and atomic Mott insulations. In Ref. [11], Zubairy and co-workers realized the phase control of the atom localization and reduced the uncertainty in a particular position measurement of the single atom by a factor of 2 in a four-level system. They further achieved sub-half- wavelength atom localization via amplitude and phase control of the absorption spectrum [13, 14]. They found two localization peaks in either of the two half- wavelength regions along the cavity axis. Very recently, they proposed a localization scheme relying on measuring the super-fluorescence radiation scattered in a standing wave field [15]. In Ref. [16], Evers et al. realized sub-half-wavelength localization by multiple simultaneous measurements. They considered objects with an internal structure consisting of a single ground state and several excited states. They showed that multiple simultaneous measurements allow both an increase in the measurement or localization precision in a single direction and the performance of multidimensional measurements or localization.

In this present article, we will give a brief review of some of our recent research work related to atomic localization via the effects of atomic coherence and quantum interferences: atom localization based on double-dark resonance effects [23, 24], sub-half-wavelength local ization via two standing-wave fields [27], as well as its application in atom nano-lithograph[28]. Then we will introduce some of our recent related work in quantum well systems[62, 63].

Atom localization based on double-dark resonance effects

Double-dark resonances have been demonstrated in a variety of four-level systems [39–50], where the probe absorption spectrum is characterized by two EIT windows, separated by a sharp absorption peak [51]. The appearance of the central narrow structure is due to the coherent interaction between the two dark states [39], which greatly enhances the Kerr nonlinear susceptibility [55]. In this section, we present our atom localization schemes based on double-dark resonance effects in two different four-level atomic systems.

A. Atom localization in a quasi-Λ system

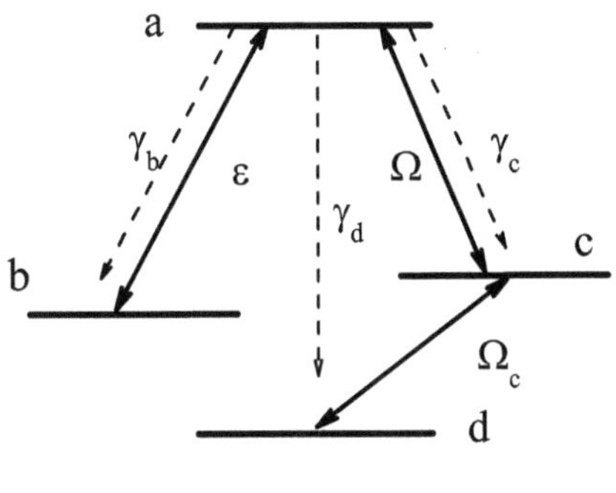

Figure1: Four-state atomic system displaying double dark resonances. γ_b, γ_c, and γ_d represents the spontaneous decay rates from the upper excited level $|a\rangle$ to the three metastable states $|b\rangle$, $|c\rangle$ and $|d\rangle$.

The atomic system under consideration is shown in Fig. 1. Here $|a\rangle$ is an upper excited state and $|b\rangle$, $|c\rangle$ and $|d\rangle$ are three lower metastable states. The transition $|a\rangle \leftrightarrow |c\rangle$ is taken to be nearly resonant with a driving field with Rabi frequency $\Omega = \Omega_0 \sin(kx)$, where $k = 2\pi/\lambda$ and λ is the wavelength. A weak probe field with Rabi frequency ε couples the states $|a\rangle$ and

$|b\rangle$. Therefore, they form a conventional and simple Λ configuration. An additional coherent perturbation field with Rabi frequency Ω_c couples the state $|c\rangle$ to another state $|d\rangle$, and this leads to the occurrence of double-dark resonances [39].

In the interaction picture with the rotating-wave approximation, the density matrix equations of the system are:

$$\dot{\rho}_{aa} = -(\gamma_b + \gamma_c + \gamma_d)\rho_{aa} + i\varepsilon(\rho_{ba} - \rho_{ab}) + i\Omega(\rho_{ca} - \rho_{ac}), \tag{1a}$$

$$\dot{\rho}_{ab} = -\Gamma_{ab}\rho_{ab} + i\varepsilon(\rho_{bb} - \rho_{aa}) + i\Omega\rho_{cb}, \tag{1b}$$

$$\dot{\rho}_{ac} = -\Gamma_{ac}\rho_{ac} + i\varepsilon\rho_{bc} - i\Omega_c\rho_{ad} + i\Omega(\rho_{cc} - \rho_{aa}), \tag{1c}$$

$$\dot{\rho}_{ad} = -\Gamma_{ad}\rho_{ad} + i\varepsilon\rho_{bd} + i\Omega\rho_{cd} - i\Omega_c\rho_{ac}, \tag{1d}$$

$$\dot{\rho}_{cd} = -\Gamma_{cd}\rho_{cd} + i\Omega_c(\rho_{dd} - \rho_{cc}) + i\Omega\rho_{ad}, \tag{1e}$$

$$\dot{\rho}_{cc} = \gamma_c\rho_{aa} + i\Omega_c(\rho_{dc} - \rho_{cd}) + i\Omega(\rho_{ac} - \rho_{ca}), \tag{1f}$$

$$\dot{\rho}_{db} = -\Gamma_{db}\rho_{db} - i\varepsilon\rho_{da} + i\Omega_c\rho_{cb}, \tag{1g}$$

$$\dot{\rho}_{dd} = \gamma_d\rho_{aa} + i\Omega_c(\rho_{cd} - \rho_{dc}), \tag{1h}$$

$$\dot{\rho}_{bb} = \gamma_b\rho_{aa} + i\varepsilon(\rho_{ab} - \rho_{ba}), \tag{1i}$$

$$\dot{\rho}_{bc} = -\Gamma_{bc}\rho_{bc} + i\varepsilon\rho_{ac} - i\Omega\rho_{ba} - i\Omega_c\rho_{bd}, \tag{1j}$$

with $\rho_{ij} = \rho^*_{ji}$ and the closure relation $\sum_i \rho_{ii} = 1$ ($i, j = a,b,c,d$). Here Here γ_b, γ_c, and γ_d represent the spontaneous decay rates from state $|a\rangle$ to states $|b\rangle$, $|c\rangle$, and $|d\rangle$, respectively. Here $\Gamma_{ab} = \gamma_{ab} - i\Delta$, $\Gamma_{ac} = \gamma_{ac} - i\Delta_0$, $\Gamma_{ad} = \gamma_{ad} - i(\Delta_0 + \Delta_c)$, $\Gamma_{cd} = \gamma_{cd} - i\Delta_c$, $\Gamma_{bd} = \gamma_{bd} - i(\Delta - \Delta_c - \Delta_0)$, $\Gamma_{bc} = \gamma_{bc} - i(\Delta_0 - \Delta)$, and $\Gamma_{ij} = \Gamma^*_{ji}$ and γ_{ij} ($i \neq j$) are the relaxation rates of the respective coherence. In the nonraditive limit, $\gamma_{ab} = \gamma_{ac} = \gamma_{ad} = (\gamma_b + \gamma_c + \gamma_d)/2$ and $\gamma_{cd} = \gamma_{db} = \gamma_{bc} = 0$ and $\Delta = \nu_p - \omega_{ab}$ ($\Delta_0 = \nu - \omega_{ac}$, $\Delta = \nu_c - \omega_{cd}$) is the detuning of the probe (driving, coherent perturbation) field with frequency ν_p (ν and ν_c) and ω_{ij} ($i \neq j$) is the atomic transition frequency between levels $|i\rangle$ and $|j\rangle$.

Assuming $\varepsilon \ll \Omega, \Omega_c$, and $\rho_{bb}^{(0)} \approx 1$ ($\rho_{bb}^{(0)}$ is the initial population), then in the long-time limit, the conditional position probability distribution [8] -i.e., the probability of finding the atom in the internal excited state $|a\rangle$ and at position x in the standing-wave field is determined by a filter function [8,9,10,11,18,25] $F(x)$. In our quasi-Λ atomic system, the filter function takes the form

$$F(x) = \varepsilon^2 A^2 / [(B - \Delta A)^2 + \gamma_{ab}^2 A^2]. \tag{2}$$

Here $A = \Omega_c^2 - (\Delta_0 - \Delta)(\Delta_0 + \Delta_c - \Delta)$, $B = \Omega^2(\Delta_0 + \Delta_c - \Delta)$. Equation (2) shows that the conditional position probability distribution depends not only on three controllable detunings Δ, Δ_0, and Δ_c, but also on Rabi frequencies of the driving and coherent perturbation fields Ω and Ω_c. It is worthwhile to point out that for the case of $\Delta_0 + \Delta_c - \Delta = 0$, i.e., satisfying the three-photon resonance, the filter function is a constant for fixed ε and γ_{ab}, and not dependent on the spatial position x; this means that atom cannot be localized. Therefore, $\Delta_0 + \Delta_c - \Delta \neq 0$ is a fundamental condition for realizing atom localization in our scheme.

We would like to find out the effect of the probe detuning on the atom localization. Because of the periodicity of the standing-wave field, we only study the variance of $F(x)$ in the subwavelength domain($-\pi \leq kx \leq \pi$). For the convenience of discussion and without lost of generality, we choose $\gamma_b = \gamma_c = \gamma_d = \gamma = 1.0$. Considering the driving and coherent perturbation fields on resonance with the corresponding transitions-i.e., $\Delta_0 = 0$ and $\Delta_c = 0$ and setting $\Omega_0 = 10\gamma$ and $\Omega_c = 0.2\gamma$, we present a three-dimensional (3D) demonstration of the conditional position probability distribution $F(x)$ as a function of kx and the probe-field detuning Δ, shown in Fig. 2(a). Figure 2(a) shows that the probe-field detuning has a significant effect on the atom localization. When Δ is zero, the conditional probability distribution is space invariant; therefore, there does not exist any atom localization. This has been clarified for the case of three-photon resonance in the above. When the probe field has a little deviation from resonance, $\Delta \neq 0$, some distribution peaks corresponding to different positions occur. Therefore, atom localization is realized. Moreover, it is noteworthy that these position probability peaks just lie at the nodes of the standing-wave field; that is, the atom is localized just at the nodes of the standing-wave field. On average, only two sharp peaks occur in the subwavelength domain. In the conventional scheme

[8,9,10,18], for a single required frequency measurement, the probability of finding the atom at a particular position is $1/4$. While in our present scheme the existence of the two equally probable sharp peaks means that the detecting probability of this atom in the subwavelength domain increases to $1/2$. This can significantly improve the measurement precision for this atom. With a further increment of Δ, there gradually occur four sharp peaks of the atom localization in the subwavelength domain. This means that the atomic detecting probability is again reduced to $1/4$. We can say that adjusting the probe-field detuning can realize the quantum control of atom localization and reduce the uncertainty in measuring a particular position of the single atom by a factor of 2.

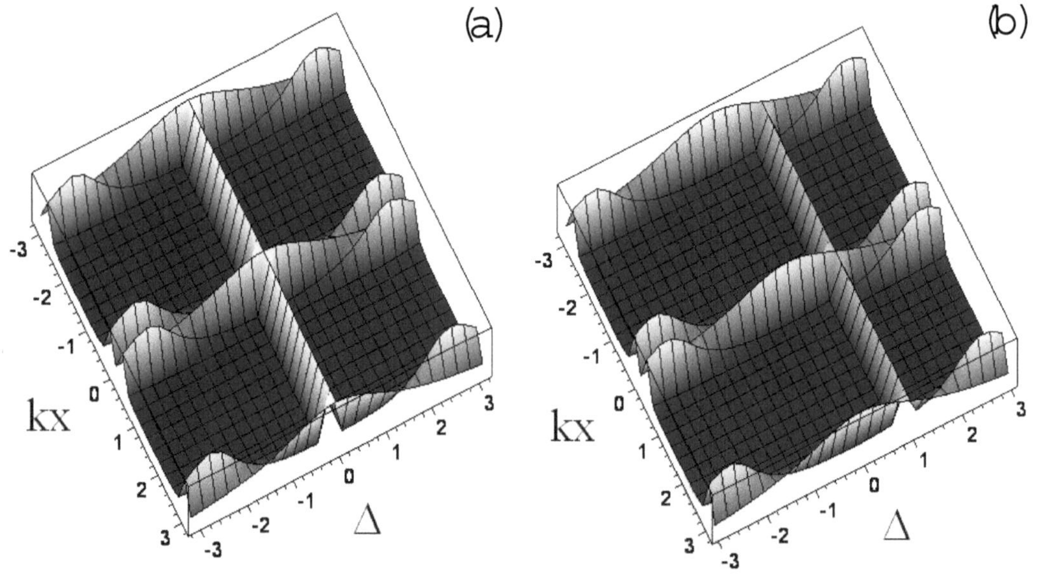

Figure 2: The conditional position probability distribution $F(x)$ (in arbitrary units) as a function of kx and Δ. (a) $\Delta_c = 0.0$, (b) $\Delta_c = 1.0\gamma$. Other parameters are $\Delta_0 = 0.0$, $\Omega_0 = 10\gamma$ and $\Omega_c = 0.2\gamma$.

This can be explained from our four-level atomic system model: when the additional perturbation field is added, there exist two dark states, and their mutual interference can induce a sharp light absorption [39]. In the case around the three-photon resonance, the two dark-state resonances play nearly equal roles and their interference induces the occurrence of three sharp peaks of atom localization in the subwavelength domain. When the probe-field detuning is large, the role of one of the two dark states is weakened while the other is built up. In the limit of $|\Delta| \to \infty$, only one dark state is dominant and this is similar to that in a Λ-type system [18]. Therefore, there exist four sharp peaks of atom localization in the subwavelength domain.

Figure 2(a) also shows that, when both the coherent perturbation and driving fields are resonant, the spatial variance of atom localization is even symmetric about the zero-probefield detuning, $\Delta = 0$. When one of the two fields is nonresonant, this symmetry will be destroyed due to the loss of symmetry in double-dark states [39]. As an example, Fig. 2(b) illustrates the case that the coherent perturbation field is nonresonant.

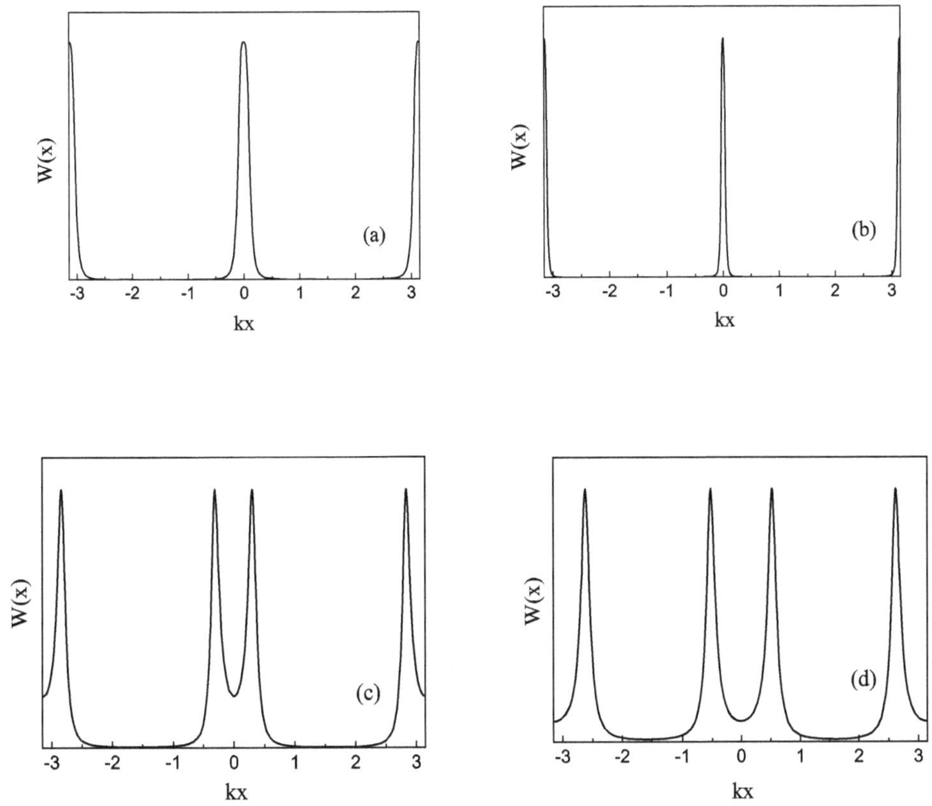

Figure 3: The conditional position probability distribution $F(x)$ (in arbitrary units) as a function of kx for four different probe-field detunings, (a) $\Delta = 0.05\gamma$, (b) $\Delta = 0.15\gamma$, (c) $\Delta = 3.0\gamma$, and (d) $\Delta = 5.0\gamma$. All other parameters are the same as in Fig. 2.

In order to see more clearly the effect of probe-field detuning on atom localization, in Fig. 3 we present a 2D demonstration of the conditional position probability distribution $F(x)$ as a function of kx for four different probe-field detunings. Figure 3 shows that probe-field detuning has a significant effect not only on the numbers of atom localization peaks, but also on the degree of atom localization. When Δ is small [Figs. 3(a) and 3(b)], there exist three peaks in the subwavelength domain. The peaks in Fig. 3(a) are more pronounced than those in Fig. 3(b). This indicates that a suitable increment in the probe-field detuning can improve the degree of atom localization, which is different from that in the scheme proposed by Paspalakis and Knight [18] where the smaller the probe detuning is, the more pronounced the localization peak is. When the probe-field detuning is large [Figs. 3(c) and 3(d)], the degree of atom localization becomes worse and the signal-to-noise ratio is larger than those in Figs. 3(a) and 3(b).

B. Atom localization in a tripod system

Among various four-level systems, tripod configuration has been extensively studied in the past few years. Paspalakis and Knight [48,49,50] have used a coherently prepared tripod scheme for efficient nonlinear frequency generation of new laser fields and showed that the absorption spectrum of a tripod system can be either symmetric or asymmetric by choosing appropriate parameters. Goren et al. [51] presented a method using a tripod atomic system to obtain sub-Doppler and subnatural narrowing of an absorption line that appears exactly at line center, flanked by two EIT windows.

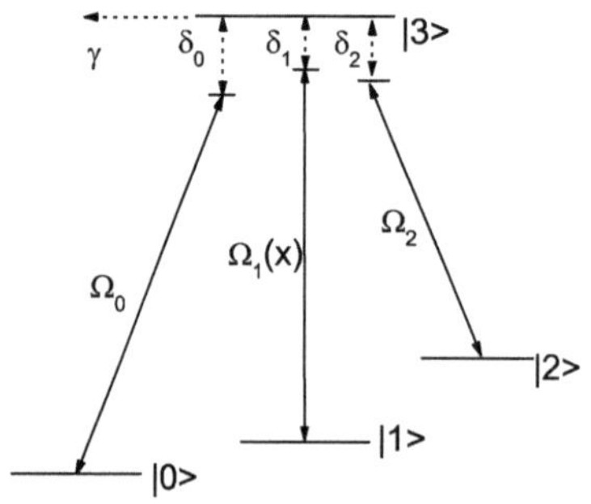

Figure 4: Schematic of the system under consideration. The atom interacts with a standing-wave field that couples the $|3\rangle \leftrightarrow |1\rangle$ transition, a probe laser field that couples the $|3\rangle \leftrightarrow |0\rangle$ transition, and a control laser field that couples the $|3\rangle \leftrightarrow |2\rangle$ transition.

Now we propose an atom localization scheme in a tripod atomic system, as Fig. 4 shows. $|0\rangle$, $|1\rangle$ and $|2\rangle$ are a lower level and two metastable levels respectively. A single upper level $|3\rangle$ decays out of the system. The atom, moving in the z direction, passes through the classical standing-wave field aligned along the x axis. The transition $|3\rangle \leftrightarrow |0\rangle$ is coupled by a weak probe field with Rabi frequency Ω_0. The transition $|3\rangle \leftrightarrow |1\rangle$ is coupled by a classical standing-wave field with position-dependent Rabi frequency $\Omega_1(x) = \Omega_1 sin(kx)$. The transition $|3\rangle \leftrightarrow |2\rangle$ is coupled by an additional control field with Rabi frequency Ω_2.

We consider that the atom is initially in its ground state $|0\rangle$ and that the probe field is very weak. We restrict our discussion under the limitation that $\delta_1 = -\delta_2$. Moreover, we applied an exactly resonant probe field $\delta_0 = 0$. The conditional position probability distribution is determined by the filter function

$$F(x)=\frac{\Omega_0^2}{\left[\frac{\Omega_2^2-\Omega_1^2\sin(kx)}{\delta_2}\right]^2+\frac{\gamma^2}{4}}. \tag{3}$$

The dependence of $F(x)$ on x makes it possible to obtain information about the x position of the atom as it passes through the standing-wave field via measuring the population in upper state, which may be realized in the laboratory using a standard spectroscopic method[7,56], the fluorescence shelving techniques[57], or the heterodyne measurement of fluorescence from a single atom[58,59].

It is easy to find from Eq. (3) that the maxima of the peaks are located at

$$kx=\pm\arcsin(\Omega_2/\Omega_1)+m\pi. \tag{4}$$

where m is an integer. The FWHM of all the peaks is given by

$$w=\left|\arcsin\frac{\sqrt{\Omega_2^2+\gamma\delta_2 2}}{\Omega_1}-\arcsin\frac{\sqrt{\Omega_2^2-\gamma\delta_2 2}}{\Omega_1}\right|. \tag{5}$$

From Eqs. (3)-(5), it is easy to see that the behavior of atom localization can be manipulated by parameters of the additional control field. This feature reflects the idea that applying an additional field to disturb the original dark state may produce double-dark resonances, and that the interaction between the double-dark states can be engineered by the parameters of the additional control field.

From Eqs. (3)-(5), we find that for given Ω_1 and Ω_2, the peak position is fixed, but the peak width can still be narrowed by decreasing δ_2. In other words, the localization resolution can be enhanced by decreasing δ_2. However, when δ_2 is decreased exactly to be zero, i.e., the two transparency windows coincide at $\delta_0=0$, then perfect EIT occurs at $\delta_0=0$ instead of absorption [49,51]. In such a case, the atom localization cannot be realized. This can also be seen from Eq. (3), $F(x)\equiv 0$ for $\delta_2=0$. On the contrary, if the frequency detunings are even slightly tuned to be nonzero, the localization precision is greatly enhanced. What is more, from Eq. (4), we can find that for $\Omega_2\in(0,\Omega_1)$, there are four probable peak positions within the interval $(-\pi,\pi)$ and the probability of finding the atom at a particular position for a single population detection is always $1/4$. When $\Omega_2=\Omega_1$, the number of localization peaks is reduced to two. The numerical results are shown in Figs. 5(a)-(d).

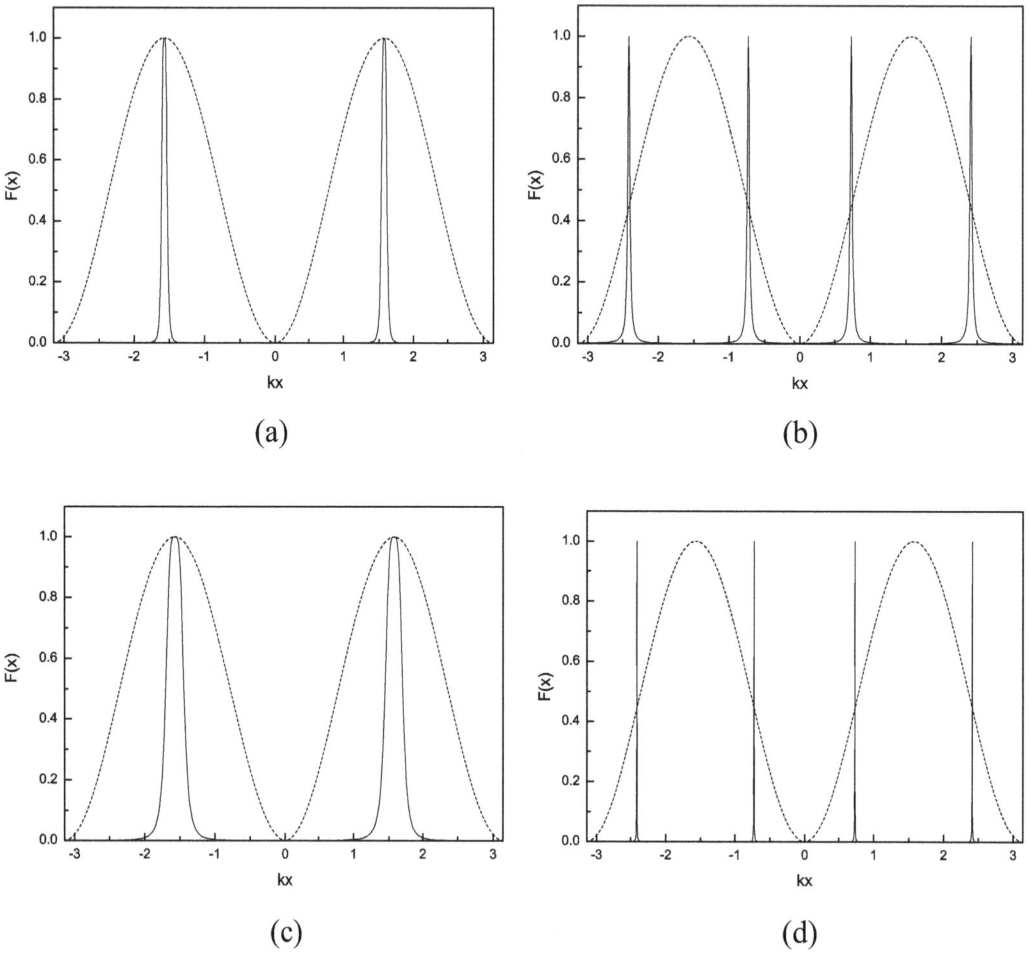

Figure 5: Filter function $F(x)$ (solid curve) as a function of kx. The dashed curve is a sine-squared function. (a) $\Omega_2 = 2.00$, $\delta_2 = -\delta_1 = 1.50$, (b) $\Omega_2 = 3.00$, $\delta_2 = -\delta_1 = 1.50$, (c) $\Omega_2 = 2.00$, $\delta_2 = -\delta_1 = 0.15$, (d) $\Omega_2 = 3.00$, $\delta_2 = -\delta_1 = 0.15$. Other parameters are $\delta_0 = 0.00$, $\Omega_1 = 3.00$, and $\gamma = 0.20$. All parameters are measured in arbitrary units.

From the above analysis, we can see that the interacting double-dark resonances provide us the opportunity to control the localization results by the parameters of the control field.

Sub-half-wavelength localization via two standing-wave fields

In all above studies, only one standing-wave field is applied to the atomic system. What will happen if the atom is driven by more than one standing-wave fields? In order to figure this out, we investigate localization behaviors in a four-level ladder-type atomic system (shown in Fig. 6), where two classical standing-wave fields s1 and s2 are applied to couple the two upper transitions $|3\rangle \leftrightarrow |2\rangle$ and $|4\rangle \leftrightarrow |3\rangle$, respectively. Their Rabi frequencies are dependent on the position and are defined by $\Omega_{si}(x) = \Omega_{si}\sin(k_i x)$ ($i = 1,2$). These two standing-wave fields may coexist in the cavity as long as the condition $L = m\pi/k_1 = n\pi/k_2$ is satisfied, where L is the length of the cavity, and m, n are integers. In order to investigate the position and width of the localization peaks, we apply a weak probe field with the Rabi frequency Ω_p to couple the transition $|2\rangle \leftrightarrow |1\rangle$. We consider an atom, moving in the z direction, as it passes through the classical standing-wave fields of a cavity. The optical fields are along the x - y plane, so that the atom meets them at the same time as it travels along the z direction. The cavity is taken to be aligned along the x axis, and the probe field can be applied in the y direction. The proposed system can be realized in ^{85}Rb atoms. We denote the ratio of the wave vectors of the two standing waves as $p = k_2/k_1$ which is determined by the nature of atoms. For example, in ^{85}Rb atomic system, $p \in (0.508, 0.586)$ and in the numerical calculation, we choose $p = 0.55$.

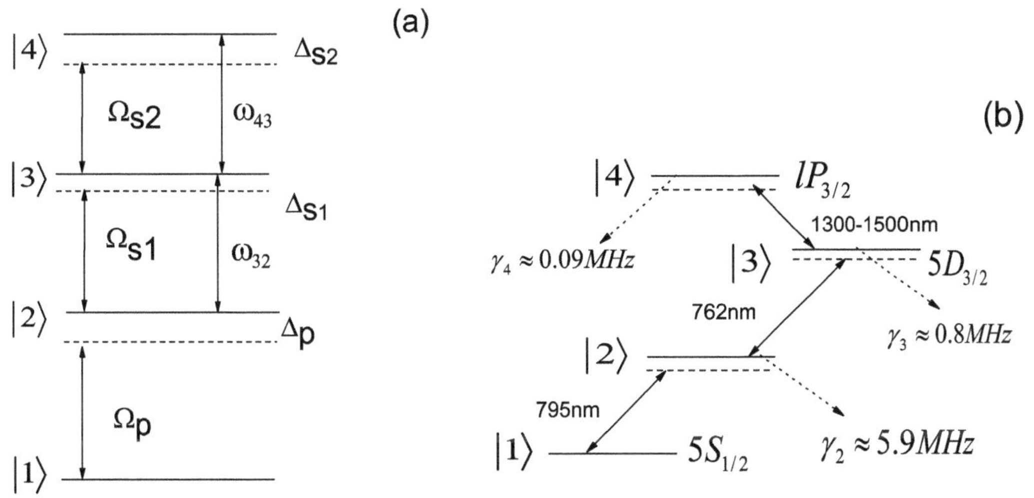

Figure 6: (a) Energy level structure for consideration. (b) Specific case for ^{85}Rb atom as an example.

As the probe field Ω_p is sufficiently small as compared to the standing wave Rabi frequencies Ω_{s1} and Ω_{s2}, we solve the density matrix equations in steady state in the nondepletion approximation ($\rho_{11} \doteq 1$). Furthermore, we consider the situation that the standing-wave fields are both exactly resonant with corresponding transitions. The imaginary part of the susceptibility, which accounts for absorption, takes the form

$$\frac{\chi''}{N} = Im(\frac{\rho_{21}}{\Omega_p}) = \frac{\left(4\Delta_p^2 - \Omega_{s2}^2 \sin^2 k_2 x\right)^2}{\left(4\Delta_p^2 - \Omega_{s2}^2 \sin^2 k_2 x\right)^2 + 4\Delta_p^2\left(4\Delta_p^2 - \Omega_{s1}^2 \sin^2 k_1 x - \Omega_{s2}^2 \sin^2 k_2 x\right)^2}, \quad (6)$$

where $N = 2N|\wp 21|^2 / \varepsilon_0 \hbar$, and N is the atom number density in the medium. From Eq. (6), we can see that χ'' is dependent on position x, the probe detuning Δ_p, as well as the Rabi frequencies Ω_{s1} and Ω_{s2}. It is, in principle, possible to obtain information about the x position of the atom as it passes through the cavity. Thus the measurement of position-dependent quantities can provide information on the atomic position. By adjusting these parameters, we can control the atom localization behavior, to increase the detecting probability and to enhance the localization precision.

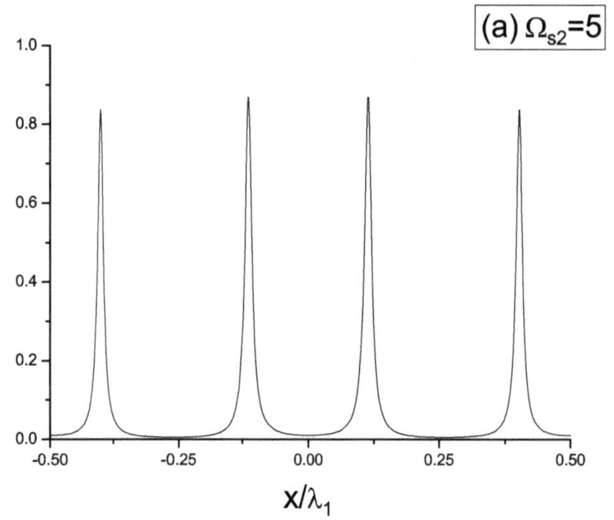

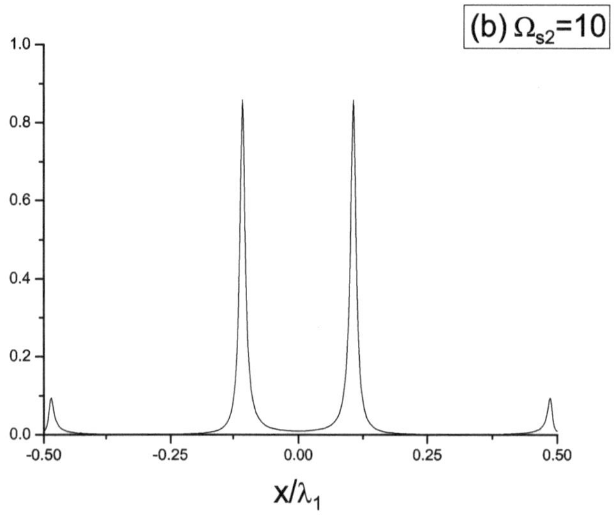

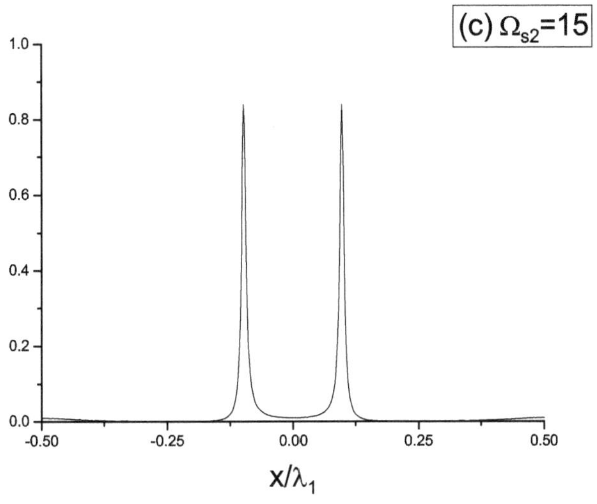

Figure 7: The probe absorption when two standing-wave fields are applied. (a) $\Omega_{s2} = 5.0$, (b) $\Omega_{s2} = 10.0$, (c) $\Omega_{s2} = 15.0$. The common parameters are $p = 0.05$, $\Delta_p = 5.0$, $\Omega_{s1} = 15.0$, $\gamma_3 = 0.14$, and $\gamma_4 = 0.015$.

In Fig. 7 we plot the normalized imaginary part of the susceptibility at the probe frequency as a function of the normalized position x with different Ω_{s2}. The parameters are chosen as $\Delta_p = 5.0$ and $\Omega_{s1} = 15.0$. In Fig. 7(a), it can be seen that there are four peaks in the subwavelength domain, as usual, and the probability of finding the atom at every possible position is 1/4. We then increase Ω_{s2} into $\Omega_{s2} = 10.0$, and find that two of the four localization peaks are lowered [see Fig. 7(b)]. When the Rabi frequency of the classical standing-wave field s2 is increased much more, for example, $\Omega_{s2} = 15.0$, those two peaks near $x = \pm\lambda_1/2$ almost disappear, and the detecting probability increased from 1/4 to 1/2 [shown in Fig. 7(c)]. More importantly, we find that the two localization peaks remained just occur in the half-wavelength domain, i.e., $-\lambda_1/4 < x < \lambda_1/4$. We find that the second standing-wave field plays an important role in atom localization. By increasing Ω_{s2}, we can restrain certain localization peaks and confine the atom to half-wavelength regions, which is called sub-half-wavelength localization.

Then we turn to the localization precision problem, which is shown by the width of the localization peak. In Fig. 8, we plot the normalized imaginary part of the susceptibility versus the normalized position x under the condition that sub-half-wavelength localization has occurred.

We choose the parameters as $\Delta_p = 5.0$, $\Omega_{s2} = 15.0$, and then increase the Rabi frequency of the first standing-wave field Ω_{s1} (see the solid and dashed curve in Fig. 8). We find that, increasing Ω_{s1} makes the localization peaks narrowed down. The localization precision can be much improved. At the same time, the number of localization peaks does not change, i.e., the probability of finding the atom within a half wavelength domain of the standing-wave field remains unchanged. This suggests, by increasing the intensity of the classical standing-wave field s1, that the localization precision is enhanced.

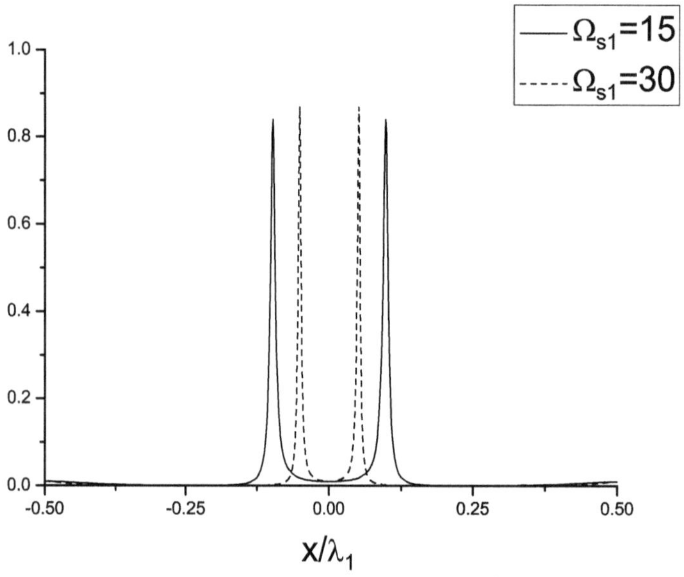

Figure 8: The probe absorption when two standing-wave fields are applied. The solid curve: $\Omega_{s1} = 15.0$, and the dashed curve: $\Omega_{s1} = 30.0$. The common parameters are $p = 0.05$, $\Delta_p = 5.0$, $\Omega_{s2} = 15.0$, $\gamma_3 = 0.14$, and $\gamma_4 = 0.015$.

Application: atom nano-lithograph scheme via two orthogonal standing-wave fields

Inspired by our previous work with two standing-wave fields in a ladder-type system, we suggest a scheme for achieving 2D atom nano-lithograph based on 2D atom localization, using two cavities orthogonal to each other. The atomic system is the same as that illustrated in Fig. 6. But, instead of applying two standing-wave fields in the same direction, here the cavities are taken to be aligned along the x and y axes, respectively, orthogonal to each other. Their Rabi frequencies are dependent on the position and are defined by $\Omega_{s1}(x) = \Omega_{s1}\sin(k_{s1}x)$ and $\Omega_{s2}(y) = \Omega_{s2}\sin(k_{s2}y)$. The experimental setup is illustrated in Fig. 9.

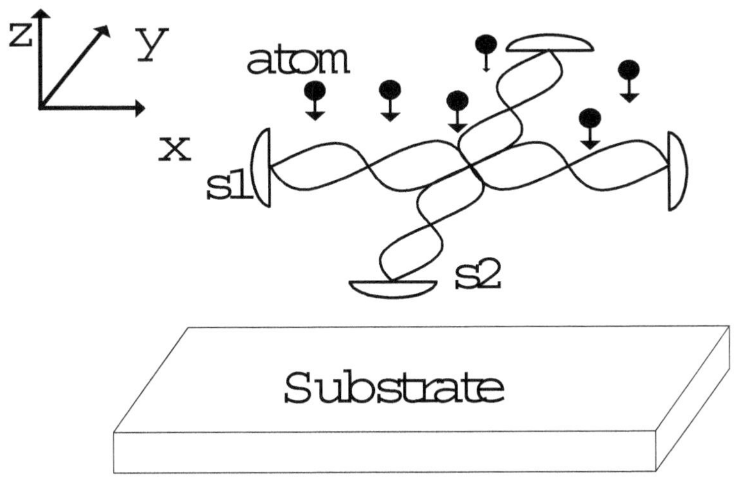

Figure 9: Schematic diagram of the setup. Two cavities are taken to be aligned along the x and y axes, respectively. The atoms travel along the z direction and then deposit on the substrate.

The probe absorption is now related on the directions x and y. The conditional position probability of finding the atom at a precise position in the standing-wave field, after the probe absorption, is then determined by the filter function

$$F(x,y;t|n) = A^2 C_2(x,y,t) C_1^*(x,y,t) = A^2 \rho_{21}(x,y,t). \tag{7}$$

Where $C_n(x,y,t)$ is the time- and position-dependent probability amplitude of the atom being in level $|n\rangle$. The atom, when passing through the standing-wave fields in the z direction, is then localized in the xy plane.

Using the same density matrix approach, we can obtain the imaginary part of the susceptibility, which accounts for absorption,

$$\frac{\chi''}{N} = Im(\rho_{21}/\Omega_p) = \frac{\left(\Omega_{s2}^2(y) - 4\Delta_p^2\right)^2}{\left(\Omega_{s2}^2(y) - 4\Delta_p^2\right)^2 + 4\Delta_p^2\left(\Omega_{s1}^2(x) + \Omega_{s2}^2(y) - 4\Delta_p^2\right)^2}. \tag{8}$$

Equation (8) exhibits that χ'' is dependent on the position (x,y), the probe detuning Δ_p, as well as the Rabi frequencies Ω_{s1} and Ω_{s2}. By adjusting the optical fields, we can control the atom localization behavior, to remodel the localization topography and to enhance the localization precision. Periodic refractive index structure can be achieved owing to the periodicity of the standing-wave fields, showing the feasibility of fabricating 2D periodic patterns via atom nano-lithography.

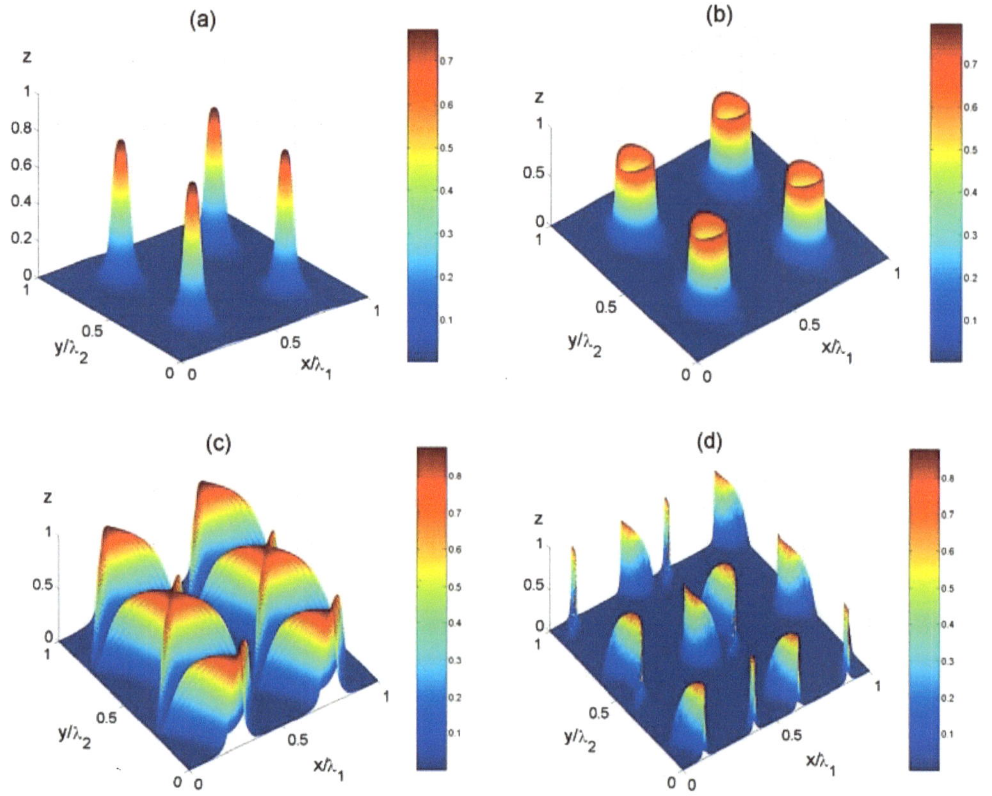

Figure 10: (Color online) Numerical simulations of the resulting patterns via 2D atom nano-lithography achieved by atom localization. (a) $\Omega_{s1,s2}=10\sqrt{2}$, (b) $\Omega_{s1,s2}=15.0$, (c) $\Omega_{s1,s2}=20.0$, and (d) $\Omega_{s1,s2}=30.0$. The common parameters are $\Delta_p=10.0$, $\gamma_3=0.14$, and $\gamma_4=0.015$.

From our numerical studies, we find that our scheme of 2D localization provide a potential application in atom nano-lithography. With a setup illustrated in Fig. 9, we can lithograph the atoms on a substrate, and fabricate periodic refractive index structures. The numerical simulations of the resulting patterns via 2D atom nano-lithography achieved by our localization method are shown in Fig. 10(a)-(d). The patterns are fabricated on a substrate in the the xy plane and grow in the z direction (in arbitrary units). In Fig. 10(a), we achieve nanorod arrays disposed on a substrate in the xy plane. The diameter of the rods is about $1/10$ of the wavelengths of the standing waves, or about several tens of nanometers. The period of the arrays is one half of the wavelengths of the standing-wave fields, e.g., a few hundred nanometers. The height of the nanorods in the z direction can be controlled by the lithographing time and the atom density: The longer it takes, or the more atoms are localized, the higher the nanorods grow, resulting in deeper structures. When the intensities of the standing-wave fields increase, the resulting pattern changes. We find that, as shown in Fig. 10(b), hollow capillary arrays can be fabricated on the substrate. The centers of the capillaries are the antinodes of the standing-wave fields. The diameter is about $0.2\lambda_{1,2}$ and the thickness is about $0.05\lambda_{1,2}$. We achieve the ‘crosses’ and ‘parentheses’ structures by further increasing the standing-wave intensities. Furthermore, we can combine these structures together. For example, we can first localize the atoms when $\Omega_{s1,s2}=10\sqrt{2}$, and prepare the nanorod arrays. Then we increase $\Omega_{s1,s2}=15$, and achieve the hollow capillary arrays, with a nanorod deposits in each center.

In addition, we have proposed some other atom localization schemes in four-level atomic systems. Phase-sensitive atom localization has been realized in a loop Λ -system[25]. Due to the sensitivity of the loop system to the relative phases between the coupled fields, the detection probability of atoms within the sub-wavelength domain of the standing wave can be improved by a factor of 2 while appropriate choosing the relative phase. In a four-level alkaline earth atomic system[26], not only the positions but also the widths of the localization peaks are investigated systemically. It is shown that the numbers and the widths of the localization peaks can be controlled by adjusting the additional control field.

Some recent related work in quantum well system

There is great interest in extending the studies of nonlinear interactions to semiconductors, not only for the understanding of the nature of quantum coherences in semiconductors, but also for the possible implementation of optical devices based on these properties. It is well known, in the conduction band of semiconductor quantum structure, that the confined electron gas exhibits atomiclike properties. For examples, strong EIT[64], tunneling induced transparency (TIT)[65], ultrafast all-optical switching[66], slow light[67], etc. More recently[68], we extended the work of Nakajima[69] into an asymmetric GaAs quantum wells (QWs), and achieved enhancement of self-Kerr nonlinearity based on Fano interference[70]. This type interference in semiconductor intersubband transitions has been observed experimentally[71], and the sign of quantum interference (constructive or destructive) in optical absorption can be reversed by varying the direction of tunneling from a double of excited energy level to a common continuum[72].

A. Tunnelling induced large XPM in an asymmetric quantum wells

We propose an asymmetric double AlGaAs/GaAs quantum well structure with a common continuum to generate a large cross-phase modulation (XPM). The basic idea is to combine resonant tunneling induced constructive interference in cross-phase modulation (XPM) and tunneling-induced transparency (TIT). The band structure is shown in Fig. 11, which is designed with small electron decay rates, which can reduce the linear absorption effectively. A $Al_{0.16}Ga_{0.84}As$ layer with thickness of 6.7 nm is separated from a 7.9 nm GaAs layer by a 4.2 nm $Al_{0.44}Ga_{0.56}Al$ potential barrier. What on the right side of the right well is a thin (2.6nm) $Al_{0.44}Ga_{0.56}As$ barrier, which is followed by a thick $Al_{0.17}Ga_{0.83}As$ layer. In this structure, one would observe the ground subband $|1\rangle$ with energy 46.0 meV and the first excited subband

$|4\rangle$ with energy 304.8 meV in the right deep and left shallow wells, respectively. Two new subbands $|2\rangle$ and $|3\rangle$ are created by mixing the ground subband of the shallow well $|sg\rangle$ and the first excited subband of the right deep well $|de\rangle$ by tunneling, and they have energies 179.2 meV and 183.1 meV, respectively. The solid lines represent the corresponding wave functions. A probe field with Rabi frequency Ω_p couples subband $|1\rangle$ to subbands $|2\rangle$ and $|3\rangle$, which are themselves coupled to the common excited subband $|4\rangle$ via a signal field with Rabi frequency Ω_s.

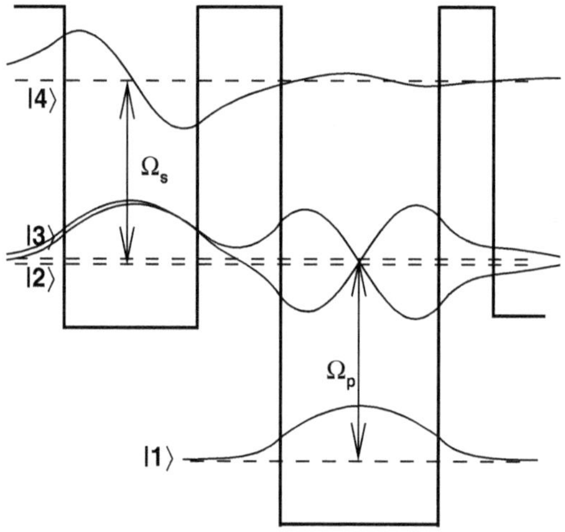

Figure 11: Conduction subband of the asymmetric double quantum well.

What we are interested in is the XPM nonlinearity between the signal and probe fields. We solve the coupled amplitude equations in steady state in the non-depletion approximation. Furthermore we assume $\Omega_p, \Omega_s \ll \Delta_p, \Delta_s, \delta$. Here $\Delta_p = \omega_p - (\omega_{21} + \omega_{31})/2$ and $\Delta_s = \omega_s - (\omega_{42} + \omega_{43})/2$ are the detunings of the probe and signal fields. $\delta = (\omega_3 - \omega_2)/2$. In this limit $\chi^{(1)}$ and $\chi^{(3)}_{\mathrm{XPM}}$ are, respectively, given by

$$\chi^{(1)} = \frac{2N|\mu_{12}|^2}{\hbar\varepsilon_0}\chi'^{(1)}, \tag{9}$$

$$\chi^{(3)}_{\mathrm{XPM}} = \frac{2N|\mu_{12}|^2|\mu_{34}|^2}{\hbar^3\varepsilon_0}\chi'^{(3)}_{\mathrm{XPM}}, \tag{10}$$

where μ_{12} and μ_{34} denote the electric dipole matrix elements, N is the electron volume density. $\chi^{'(1)}$ and $\chi^{'(3)}_{\mathrm{XPM}}$ are given by

$$\chi^{'(1)} = -\frac{(\Delta_p - \delta + i\gamma_3) + q^2(\Delta_p + \delta + i\gamma_2) + 2iq\kappa}{(\Delta_p + \delta + i\gamma_2)(\Delta_p - \delta + i\gamma_3) + \kappa^2}, \tag{11}$$

$$\chi^{'(3)}_{\mathrm{XPM}} = \frac{1}{\Delta_p + \Delta_s + i\gamma_4} \frac{(k-q)^2}{(\Delta_p + \delta + i\gamma_2)(\Delta_p - \delta + i\gamma_3) + \kappa^2}, \tag{12}$$

where $q = \mu_{13}/\mu_{12} \simeq -0.95$ and $k = \mu_{34}/\mu_{24} \simeq 2.08$ present the ratio between the relevant subband transition dipole moment. The total electron decay rate $\gamma_i = \gamma_i^{ph} + \gamma_i^{deph}$ ($i = 2,3,4$) is the sum of the population decay originate from tunnelling (γ_i^{ph}) and the dephasing rate (γ_i^{deph}) which are introduced to account not only for intrasubband phonon scattering and electron-electron scattering, but also for inhomogeneous broadening due to scattering on interface roughness. he strength of the cross coupling is assessed by $\eta = \kappa/(\gamma_2\gamma_3)^{1/2}$, which can be augmented by decreasing the temperature which generally leads to smaller dephasing γ_i^{deph}.

It is worth noting that the key feature of our structure is, owing to resonant tunnelling, that the wave functions of subbands $|2\rangle$ and $|3\rangle$ are symmetric and antisymmetric combinations (that is, quantum mechanically anticross) of $|sg\rangle$ and $|de\rangle$ (see Fig. 11). As a result, $q \cdot k < 0$, which indicates constructive interference in XPM nonlinearity [see Eq. (12)].

$\mathrm{Im}[\chi^{'(1)}]$, $\mathrm{Im}[\chi^{'(3)}_{\mathrm{XPM}}]$ and $\mathrm{Re}[\chi^{'(3)}_{\mathrm{XPM}}]$, which, respectively, account for the linear absorption, two-photon absorption and XPM. In Fig. 12, we plot their evolutions versus the probe photon energy with $\eta = 0.83$ and $\Delta_s = -0.55$ meV. It is clear, within the transparency window, that the strengths of XPM and two-photon absorption are enhanced. While, the two-photon absorption peak is very sharp. Therefore, for certain probe detuning, for example, at dot A, linear and two-photon absorptions are vanished. While, very fortunately, the strength of XPM is large. $\mathrm{Re}\,[\chi^{'(3)}_{\mathrm{XPM}}] \simeq -9.6$ meV $^{-3}$ ($\mathrm{Re}\,[\chi^{(3)}_{\mathrm{XPM}}] \simeq -7.3\times10^{-7}$ $\mathrm{m^2V^{-2}}$). This suggests that, in our structure, large XPM can be achieved with vanishing linear and two-photon absorptions simultaneously. This interesting result is produced by the combination of a destructive interference in linear absorption with a constructive interference in the nonlinear susceptibility associated with XPM, which is induced by resonant tunnelling. Our numerical analysis show that the strength of XPM can be enhanced much more by increasing the strength of Fano interference, and at the same time, a further reduction is desirable, and will lead to improved transparency.

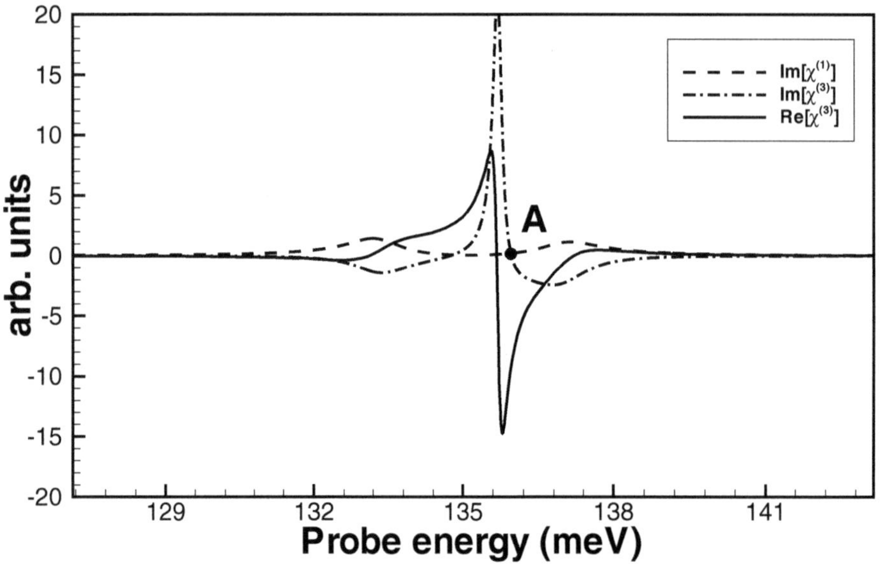

Figure 12: The evolutions of $\mathrm{Im}[\chi^{'(1)}]$ (dashed line), $\mathrm{Im}[\chi^{'(3)}_{\mathrm{XPM}}]$ (dash-dot line) and $\mathrm{Re}[\chi^{'(3)}_{\mathrm{XPM}}]$ (solid line) versus the probe energy with $\eta = 0.83$.

One prominent application of large XPM in our structure would be the creation of an all-optical switching[73] based on semiconductor material. For two matched Gaussian pulses, the nonlinear phase shift can be determined from the slowly varying envelope equation[74]. With the assumption of $\gamma_2 \approx \gamma_3$ and with the definition of the absorption cross section $\sigma_s = \omega_s \,|\, \mu_{34} \,|^2 / (\hbar \varepsilon_0 c \gamma_4)$, one obtains

$$\Phi_{\mathrm{XPM}} = n_s \frac{\sigma_s}{A} \gamma_4 \delta^2 \mathrm{Re}\left[\chi^{'(3)}_{\mathrm{XPM}}\right], \tag{13}$$

where n_s is the photon number of the signal pulse. Numerical findings indicate that our semiconductor quantum wells structure (dot A) would produce π XPM phase shift with focusing 300 photons into an area 1 $\mu\mathrm{m}^2$ (the intensity of 300 photons per nanosecond on area of 1 $\mu\mathrm{m}^2$ is ~ 600 $\mathrm{mW/cm^2}$), which has the potential application in all-optical switching.

B. Self-induced transmission on intersubband resonance in multiple quantum wells

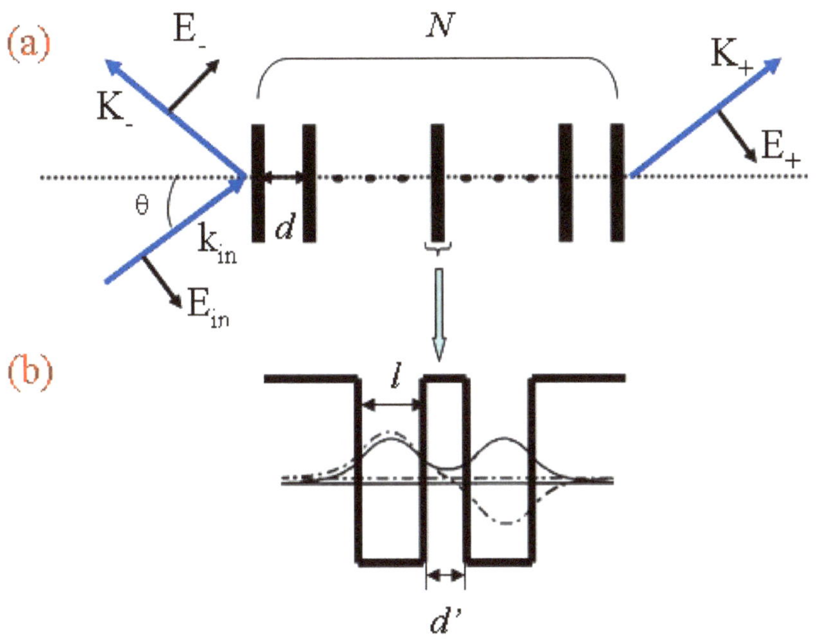

Figure 13: Schematic picture of the model configuration, which is excited by a p-polarized external electric field E_{in} at an angle of incidence θ. E_+ and E_- denote the right and left propagating total outgoing fields, respectively. The solid and dotted lines represent the lowest and excited subbands, respectively.

In this part, we show how ultrashort pulses on resonant intersubband (IS) transitions propagate nonlinearly in multiple symmetric double quantum wells (QWs). A n-type modulation-doped multiple QWs sample consisting of N are equally spaced electronically uncoupled symmetric double semiconductor GaAs/AlGaAs QWs with separation d, as shown in Fig. 13. There are only two lower energy subbands contribute to the system dynamics, $n=0$ for the lowest subband with even parity and $n=1$ for the excited subband with odd parity. The Fermi level is below the $n=1$ subband minimum, so the excited subband is initially empty. This is succeeded by a proper choice of the electron sheet density. The nonlinear propagation of ultrashort pulses on resonant IS transitions in multiple semiconductor quantum wells is described by the full Maxwell-Bloch equations:

$$\partial_t S_1(t) = [\omega_{10} - \gamma S_3(t)]S_2(t) - \frac{S_1(t)}{T_2}, \tag{14a}$$

$$\partial_t S_2(t) = -[\omega_{10} - \gamma S_3(t)]S_1(t) + 2[\frac{\wp E(t)}{\hbar} - \beta S_1(t)]S_3(t) - \frac{S_2(t)}{T_2}, \tag{14b}$$

$$\partial_t S_3(t) = -2[\frac{\wp E(t)}{\hbar} - \beta S_1(t)]S_2(t) - \frac{S_3(t)+1}{T_1}, \tag{14c}$$

$$\partial_t H_y = -\frac{1}{\mu}\partial_z E_x, \tag{14d}$$

$$\partial_t E_x = -\frac{1}{\varepsilon}\partial_z H_y - \frac{1}{\varepsilon}\partial_t P_x. \qquad (14e)$$

Here, $S_1(t)$ and $S_2(t)$ are, respectively, the mean real and imaginary parts of polarization and $S_3(t)$ is the mean population inversion per electron (difference of the occupation probabilities in the upper and lower subbands). $\wp$ is the electric dipole matrix element between the two subbands, and N_v is the electron volume density. E_x, H_y are the electric and magnetic fields, respectively. And, ε, μ are the electric permittivity and the magnetic permeability in the medium, respectively.

The structure of the multiple QWs contains $N = 50$ equally spaced double symmetric QWs with separation $d = 20\text{nm}$ $Al_{0.267}Ga_{0.733}As$ barrier, as shown in Fig. 13(a). The each pair of QWs[Fig. 13(b)] consists of two GaAs symmetric square wells of 5.5nm width and 219meV height, coupled by an $d' = 1.1\text{nm}$ $Al_{0.267}Ga_{0.733}As$ barrier as shown in Fig. 13(a). Then, the system parameters, which are dependent on the electron sheet density N_s, can be calculated[63].

The dipole moment for the structure is $\wp = -32.9e\mathring{A}$ and the relative dielectric constant is $\varepsilon_r = 12$. In addition, as the dephasing is the crucial relaxation process in semiconductor QWs, we choose $T_1 = 100\text{ps}$ and $T_2 = 10\text{ps}$ as in Ref. [75].

The full Maxwell-Bloch equations are solved by using the iterative predictor-corrector finite-diference time-domain method[76,77,78]. In what follows, we assume that the system is initially in the lowest subband. We consider a hyperbolic secant functional form for the initial incident pulse, which is described by $E_x(z=0,t) = E_0\text{sech}[1.76(t-t_0)/\tau_p]\cos[\omega_p(t-t_0)]$, where E_0 is the electric field amplitude, ω_p is the central carrier frequency. The coefficient 1.76 is used to adjust the definition of the full width at half maximum(FWHM) of the pulse intensity envelope of the laser pulse to the definition of pulse duration, in this condition, τ_p is the FWHM, and t_0 is the delay. The system we consider is excited at exact resonance, $\omega_p = \omega_{10}$, and the duration of the applied pulse is $\tau_p = 0.2\text{ps}$. For the following application, we characterize the strength of the electron-light interactions by the input pulse area, $\Theta(z=0) = \int_{-\infty}^{\infty} \frac{\wp/\sqrt{2}}{\hbar} E_x(z=0,t')dt' = \Omega_0\tau_p\pi/1.76$, where, $\Omega_0 = -(\wp/\sqrt{2})E_0/\hbar$ is the peak of the Rabi frequency.

For the larger sheet density, by varying the input pulse areas $\Theta(z=0)$, for $N_s = 5\times10^{11}\text{cm}^{-2}$ at $\Theta(z=0)=2.24\pi$ [Fig. 14(b)] and for $N_s = 8\times10^{11}\text{cm}^{-2}$ at $\Theta(z=0)=2.57\pi$ [Fig. 14(c)], the pulses transmit the structure essentially unaltered with respect to the input pulses, presuming we neglect a decrease in pulse envelope and a slight broadening induced by the dispersion. The corresponding spectra do not differ significantly from the input spectrum [Fig. 14(e) and (f)]. This phenomenon of self-induced transmission (SIT) in semiconductors do occur, which indicates that a full Rabi flopping of the electron density has occurred within the IS transitions of each double QWs.

Furthermore, we present the dependence of the input pulse area $\Theta(z=0)$ for the occurrence of self-induce transmission on the electron sheet density in Fig. 15 with the triangles. The values of N_s, which is taken up in numerical simulations, satisfies the system initial conditions that the Fermi energy is smaller than excited subband energy, i.e., the system is in the ground subband initially. It is shown that, the area of input pulse depends directly on the electron sheet density. For small sheet densities [$N_s \le 1.0\times10^{11}\text{cm}^{-2}$], the effects of electron-electron interaction are so small that the quantum-well structure equivalents to a noninteracting two-level atom system, in which 2π pulses can propagate without suffering significant losses according to the area theorem. While, the increase of the electron sheet density destroys the condition leading to SIT. The area of input pulse for the occurrence of self-induced transmission increase with the increasing electron density synchronously, and is always larger than 2π. Nevertheless, combining with the nonlinear parameters γ and β, we define an effective area of input pulse according to Ref. [75]:

$$\Theta_{eff}^2(z=0) = \Theta^2(z=0) - 2\left[(\gamma-\beta)\tau_p\pi/1.76\right]^2, \tag{15}$$

i.e., an effective coupling of light-matter. In spite the electron sheet density of QWs and the input pulse area vary in a relative large range, the effective pulse area $\Theta_{eff}(z=0)$ for the occurrence of self-induced transmission keeps invariant area 2π, as shown in Fig. 15 with squares. Considering both Fig. 14 and Fig. 15, and assuming that we neglect the decrease in the area of transmitted pulses, the phenomenon of self-induced transmission can be approximately explained by traditional standard area theorem with the effective area of input pulses.

In addition, we have also investigated the propagation properties of Gaussian pulses in the multiple QWs systems. With the same pulse duration and the central carrier frequency above, we have also found the phenomenon of self-induced transmission with Gaussian pulses. The input pulse areas $\Theta(z=0)$ for the occurrence of self-induce transmission with different electron sheet densities are shown in Fig. 15 with circles. Therefore, the feasibility for the occurrence of self-induced transmission on resonant IS transitions in multiple symmetric double semiconductor QWs can be clearly determined, which is useful to achieve a high degree of transmission for this

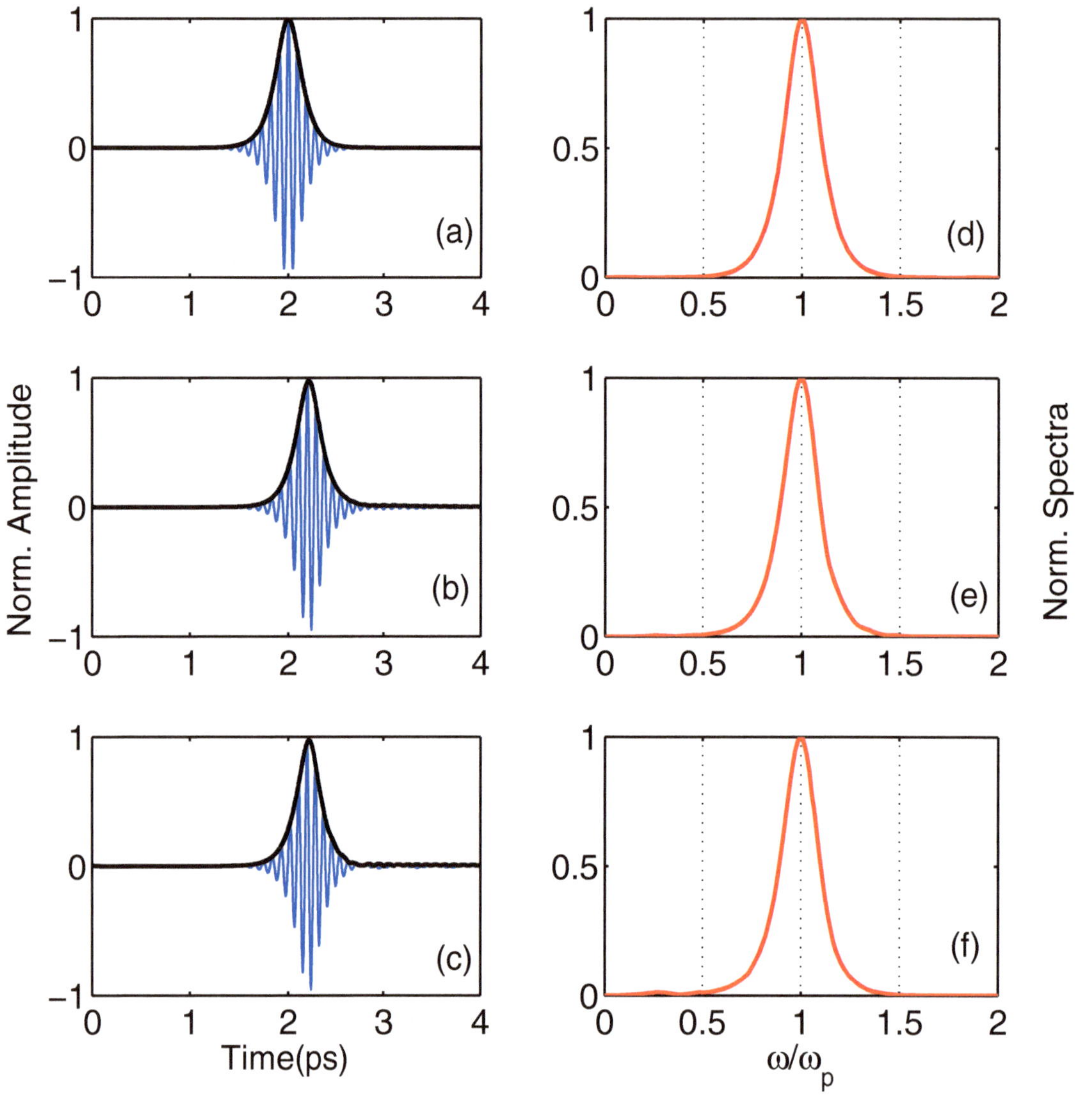

Figure 14: (Color online) Normalized amplitude (thin blue curve) of the input pulse versus time (a) and the transmitted pulses for different electron sheet densities with different input pulse area versus time, (b) $N_s = 5\times10^{11}\text{cm}^{-2}$, $\Theta(z=0) = 2.24\pi$, (c) $N_s = 8\times10^{11}\text{cm}^{-2}$, $\Theta(z=0) = 2.57\pi$. The black thick curves represent the pulse envelopes. (d)-(f) are the normalized spectra corresponding to (a)-(c), respectively.

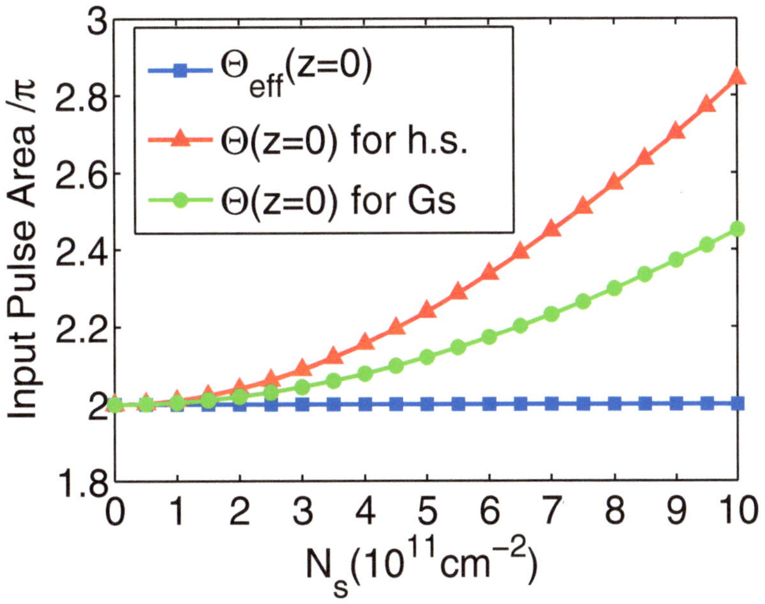

Figure 15: (Color online) The input pulse area $\Theta(z=0)$ and the effective area $\Theta_{eff}(z=0)$ (squares) for the occurrence of self-induced transmission as a function of the electron sheet densities N_s. The triangles are for hyperbolic secant pulses (h.s.) and the circles are for Gaussian pulses (Gs). The lines are a guide to the eyes.

Conclusion

In this article we reviewed some of our recent research work related to atomic localization via the effects of atomic coherence and quantum interferences. It was found that the localization property was significantly improved due to the interaction of double-dark resonances. The probability of finding the atom at a particular position could be doubled, as well as the localization precision could be dramatically enhanced. A scheme of sub-half-wavelength localization was proposed via two standing-wave fields in a ladder-type system. We also presented an application of atom localization in 2D atom nano-lithograph, through two cavities orthogonal to each other. The feasibility of control the atom localization behaviors by the external optical fields is shown. Those work may be helpful in the experimental research of atom localization.

Acknowledgements

We are sincerely grateful to Dr. C. P. Liu, D. C. Cheng and N. Cui for their previous work. This work is supported by the National Natural Sciences Foundation of China (Grant No. 60878009, 60708008, and 60978013), the Project of Academic Leaders in Shanghai (Grant No. 07XD14030), the Key Basic Research Foundation of Shanghai (Grant No.08JC1409702) and the Knowledge Innovation Program of the Chinese Academy of Sciences.

References

1. P. Storey, M. Collett, and D. Walls. Measurement-induced diffraction and interference of atoms.

Physical Review Letters 1992 Jan 27;68(4):472-475.
2. R. Quadt, M. Collett, and D. F. Walls. Measurement of Atomic Motion in a Standing Light Field by Homodyne Detection. Physical Review Letters 1995 Jan 16;74(3):351-354.
3. A. M. Herkommer, V. M. Akulin, and W. P. Schleich. Quantum demolition measurement of photon statistics by atomic beam deflection. Physical Review Letters 1992 Dec 7;69(23):3298-3301.
4. F. L Kien, G. Rempe, W. P. Schleich, and M. S. Zubairy. Atom localization via Ramsey interferometry: A coherent cavity field provides a better resolution. Physical Revew A 1997 Oct; 56(4):2972-2977.
5. J. E. Thomas. Phys. Quantum theory of atomic position measurement using optical fields. Physical Revew A 1990 Nov 1;42(9):5652-5666.
6. J. R. Gardner, M. L. Marable, G. R. Welch, and J. E. Thomas. Suboptical wavelength position measurement of moving atoms using optical fields. Physical Review Letters 1993 May 31;70(22):3404-3407.
7. H. R. Gray, R. M. Whitley, and C. R. Stroud Jr.. Coherent trapping of atomic populations. Optics Letters 1978 Dec;3(6):218-220.
8. A. M. Herkommer, W. P. Schleich, and M. S. Zubairy. Autler-Townes microscopy on a single atom. Journal of Modern Optics 1997 Nov 1;44(11-12):2507-2513.
9. S. Qamar, S. Y. Zhu, and M. S. Zubairy. Atom localization via resonance fluorescence. Physical Revew A 2000 May 11;61(6):063806(5).
10. S. Qamar, S. Y. Zhu, and M. S. Zubairy. Precision localization of single atom using Autler–Townes microscopy. Optics Communications 2000 Apr 1;176(4-6):409-416.
11. F. Ghafoor, S. Qamar, and M. S. Zubairy. Atom localization via phase and amplitude control of the driving field. Physical Revew A 2002 Apr 5;65(4):043819(8).
12. K. T. Kapale, S. Qamar, and M. S. Zubairy. Spectroscopic measurement of an atomic wave function. Physical Revew A 2003 Feb 10;67(2):023805(5).
13. M. Sahrai, H. Tajalli, K. T. Kapale, and M. S. Zubairy. Subwavelength atom localization via amplitude and phase control of the absorption spectrum. Physical Revew A 2005 Jul 25;72(1):013820(9).
14. K. T. Kapale, and M. S. Zubairy. Subwavelength atom localization via amplitude and phase control of the absorption spectrum. II. Physical Revew A 2006 Feb 15;73(2):023813(11).
15. M. Macovei, J. Evers, C. H. Keitel, and M. S. Zubairy. Localization of atomic ensembles via superfluorescence. Physical Revew A 2007 Mar 2;75(3):033801(7).
16. J. Evers, S. Qamar, and M. S. Zubairy. Atom localization and center-of-mass wave-function determination via multiple simultaneous quadrature measurements. Physical Revew A 2007 May 8; 75(5):053809(10).
17. J. T. Chang, J. Evers, M. O. Scully, and M. S. Zubairy. Measurement of the separation between atoms beyond diffraction limit. Physical Revew A 2006 Mar 10;73(3):031803(4).
18. E. Paspalakis, and P. L. Knight. Localizing an atom via quantum interference. Physical Revew A 2001 May 10;63(6):065802(4).
19. E. Paspalakis, A. F. Terzis, and P. L. Knight. Quantum interference induced sub-wavelength atomic localization. Journal of Modern Optics 2005 Aug 15;52(12):1685-1694.
20. G. S. Agarwal, and K. T. Kapale. Subwavelength atom localization via coherent population trapping. Journal of Physics B: Atomic Molecular and Optical Physics 2006 Aug 14;

39:3437-3446.
21. M. Sahrai, M. Mahmoudi, and R. Kheradmand. Atom localization of a two-level pump-probe system via the absorption spectrum. Laser Physics 2007 Jan 9;17(1):40-44.
22. J. Xu, and X. M. Hu. Localization of a two-level atom via the absorption spectrum. Physical Revew A 2006 Dec 11;364(3-4):208-213.
23. C. P. Liu, S. Q. Gong, D. C. Cheng, X. J. Fan, and Z. Z. Xu. Atom localization via interference of dark resonances. Physical Revew A 2006 Feb 3;73(2):025801(4).
24. D. C. Cheng, Y. P. Niu, and S. Q. Gong. Controllable atom localization via double-dark resonances in a tripod system . Journal of Physics B: Atomic Molecular and Optical Physics 2006 Oct 10; 23(10):2180-2184.
25. C. P. Liu, S. Q. Gong, T. Nakajima, and Z. Z. Xu. Phase-sensitive atom localization in a loop Λ-system. Journal of Modern Optics 2006 Aug 15;53(12):1791-1802.
26. L. L. Jin, H. Sun, Y. P. Niu, S. Q. Jin, and S. Q. Gong. Atom localization in a four-level alkaline earth atomic system. Journal of Modern Optics 2008 Jan 1;55(1):155-165.
27. L. L. Jin, H. Sun, Y. P. Niu, and S. Q. Gong. Sub-half-wavelength atom localization via two standing-wave fields. Journal of Physics B: Atomic Molecular and Optical Physics 2008 Apr 8; 41:085508(5).
28. L. L. Jin, H. Sun, Y. P. Niu, S. Q. Jin, and S. Q. Gong. Atom localization in a four-level alkaline earth atomic system. Journal of Modern Optics 2008 Jan 1;55(1):155-165.
29. W. D. Phillips. Nobel Lecture: Laser cooling and trapping of neutral atoms Rev. Modern Physics 1998 Jul; 70(3):721-741.
30. K. S. Johnson, J. H. Thywissen, W. H. Dekker, K. K. Berggen, A. P. Chu, R. Younkin, and et al. Localization of Metastable Atom Beams with Optical Standing Waves: Nanolithography at the Heisenberg Limit Science 1998 Jun 5;280(5369):1583-1586.
31. S. E. Harris, J. E. Field, and A. Imamoǧlu. Nonlinear optical processes using electromagnetically induced transparency. Physical Review Letters 1990 Mar 5;64(10):1107-1110.
32. K. J. Boller, A. Imamoǧlu, and S. H. Harris. Observation of electromagnetically induced transparency. Physical Review Letters 1991 May 20;66(20):2593-2596.
33. S. E. Harris. Electromagnetically Induced Transparency. Physics Today 1997 Jul;50(9): 36-42.
34. L. V. Hau, S. E. Harris, D. Dutton, and C. H. Behroozi. Light speed reduction to 17 metres per second in an ultracold atomic gas. Nature 1999 Feb 18;397(6720):594-598.
35. H. Schmidt, and A. Imamoǧlu. Giant Kerr nonlinearities obtained by electromagnetically induced transparency. Optics Letters 1996 Dec 1;21(23):1936-1938.
36. H. Wang, D. Goorskey, and M. Xiao. Enhanced Kerr Nonlinearity via Atomic Coherence in a Three-Level Atomic System. Physical Review Letters 2001 Jul 26;87(7):073601(4).
37. Y. Wu, and X. X. Yang. Highly efficient four-wave mixing in double- Λ system in ultraslow propagation regime. Physical Revew A 2004 Nov 18;70(5):053818(5).
38. Y. Wu, and L. Deng. Ultraslow Optical Solitons in a Cold Four-State Medium. Physical Review Letters 2004 Sep 28;93(14):143904(4).
39. M. D. Lukin, S. F. Yelin, M. Fleischhauer, and M. O. Scully. Quantum interference effects induced by interacting dark resonances. Physical Revew A 1999 Jan 15;60(4):3225-3228.
40. C. Y. Ye, A. S. Zibrov, Yu. V. Rostovtsev, and M. O. Scully. Unexpected Doppler-free resonance in generalized double dark states. Physical Revew A 2002 Mar 14;65(4):043805(4).
41. S. F. Yelin, V. A. Sautenkov, M. M. Kash, G. R. Welch, and M. D. Lukin. Nonlinear optics via

double dark resonances. Physical Revew A 2003 Dec 2;68(6):063801(7).

42. N. Mulchan, D. Ducreay, R. Pina, M. Yan, and Y. Zhu. Nonlinear excitation by quantum interference in a Doppler-broadened rubidium atomic system. Journal of the Optical Society of America B 2000 May;17(5):820-826.

43. G. Wasik, W. Gawlik, J. Zachorowkwski, and Z. Kowal. Competition of dark states: Optical resonances with anomalous magnetic field dependence. Physical Revew A 2001 Oct 2;64(5):051802(4).

44. E. S. Fry, M. D. Lukin, T. Walther, and G. R. Welch. Four-level atomic coherence and cw VUV lasers . Opt. Commun. 2000 May 25;179(1-6) :499-504.

45. S. R. de Echaniz, A. D. Greentree, A. V. Durrant, D. M. Segal, J. P. Marangos, and J. A. Vaccaro. Observations of a doubly driven V system probed to a fourth level in laser-cooled rubidium. Physical Revew A 2001 Jun 11;64(1):013812(8).

46. Y. F. Li, J. F. Sun, X. Y. Zhang, and Y. C. Wang. Laser-induced double-dark resonances and double-transparencies in a four-level system. Optics Communications 2002 Feb 1;202(1-3):97-102.

47. S. Q. Jin, S. Q. Gong, R. X. Li, and Z. Z. Xu. Coherent population transfer and superposition of atomic states via stimulated Raman adiabatic passage using an excited-doublet four-level atom. Physical Revew A 2004 Feb 26;69(2):023408(5).

48. E. Paspalakis and P. L. Knight. Transparency and parametric generation in a four-level system. Journal of Modern Optics 2002 Jan 15;49(1-2):87-95.

49. E. Paspalakis and P. L. Knight. Transparency, slow light and enhanced nonlinear optics in a four-level scheme. Journal of Optics B: Quantum and Semiclassical Optics 2002 Jul 29;4:372-375.

50. E. Paspalakis and P. L. Knight. Electromagnetically induced transparency and controlled group velocity in a multilevel system. Physical Revew A 2002 Jul 24;66(1):015802(4).

51. C. Goren, A. D. Wilson-Gordon, M. Rosenbluh, and H. Friedmann. Sub-Doppler and subnatural narrowing of an absorption line induced by interacting dark resonances in a tripod system. Physical Revew A 2004 Jun 2;69(6):063802(5).

52. P. Meystre, and M. Sargent . Elements of Quantum Optics 3rd edn . Berlin: Springer ;1999.

53. N. Bloembergen, and Y. R. Shen. Quantum-Theoretical Comparison of Nonlinear Susceptibilities in Parametric Media,Lasers,and RamanLasers. Physical Revew 1964 Jan 6;133(1A):A37-A49.

54. S. Rebić, D. Vitali, C. Ottaviani, P. Tombesi, M. Artoni, F. Cataliotti, and et al. Polarization phase gate with a tripod atomic system. Physical Revew A 2004 Sep 21;70(3):032317(8).

55. Y. P. Niu, S. Q. Gong, R. X. Li, Z. Z. Xu, and X. Y. Liang. Giant Kerr nonlinearity induced by interacting dark resonances. Optics Letters 2005 Dec 15;30(24):3371-3373.

56. G. Alzetta, A. Gozzini, L. Moi and G. Orriols. An experimental method for the observation of the RF transition and laser beat resonance in oriented Na vapour. Nuovo Cimento 1976; B36:5-20.

57. R. Blatt and P. Zoller. Quantum jumps in atomic systems. European Journal of Physics 1988 Jan 25;9:250-256.

58. J. T. Höffges, H. W. Baldauf, T. Eichler, S. R. Helmfrid, and H. Walther. Heterodyne measurement of the fluorescent radiation of a single trapped ion. Optics Communications 1997 Jan 1;133(1-6):170-174.

59. J. T. Höffges, H. W. Baldauf, W. Lange, and H. Walther. Heterodyne measurement of the resonance fluorescence of a single ion. Journal of Modern Optics 1997 Oct 1;44(10):1999-2010.

60. J. H. Wu and J. Y. Gao. Phase and amplitude control of the inversionless gain in a microwave-driven Λ-type atomic system . The European Physical Journal D 2003 Mar 4;23(2):315-319.
61. T. Hong, C. Cramer, W. Nagourney et al. Optical Clocks Based on Ultranarrow Three-Photon Resonances in Alkaline Earth Atoms. Physical Review Letters 2005 Feb 10;94(5):050801(4).
62. H. Sun, Y. P. Niu, R. X. Li, S. Q. Jin, and S. Q. Gong. Tunneling-induced large cross-phase modulation in an asymmetric quantum well. Optics Letters 2007 Sep 1;32(17):2475-2477.
63. N. Cui, Y. P. Niu, H. Sun, and S. Q. Gong. Self-induced transmission on intersubband resonance in multiple quantum wells. Physical Revew B 2008 Aug 26;78(7):075323(6).
64. M. C. Phillips and Hailin Wang. Electromagnetically Induced Transparency in Semiconductors via Biexciton Coherence. Physical Review Letters 2003 Oct 31; 91(18):183602(4).
65. H. Schmidt, K. L. Campman, A. C. Gossard, and A. Imamoglu. Tunneling induced transparency: Fano interference in intersubband transitions. Applied Physics Letters 1997 Jun 23; 70(25):3455 -3457.
66. H. Wu, J. Y. Gao, J. H. Xu, L. Silvestri, M. Artoni, and et al. Ultrafast All Optical Switching via Tunable Fano Interference . Physical Review Letters 2005 Jul 25;95(5):057401(4).
67. Chunhua Yuan and Kadi Zhu. Room temperature defect related electroluminescence from ZnO homojunctions grown by ultrasonic spray pyrolysis. Applied Physics Letters 2006 Aug 3; 89(5):052113(3).
68. H. Sun, S. Q. Gong, Y. P. Niu, S. Q. Jin, R. X. Li, and Z. Z. Xu. Enhancing Kerr nonlinearity in an asymmetric double quantum well via Fano interference. Physical Revew B 2006 Oct 16;74(15):155314(5).
69. T. Nakajima. Enhanced nonlinearity and transparency via autoionizing resonance. Optics Letters 2000 Jun 1;25(11):847-849.
70. U. Fano. Effects of Configuration Interaction on Intensities and Phase Shifts. Physical Revew 1961 Dec 15;124(6) :1866-1878.
71. J. Faist, C. Sirtori, F. Capassom, S.-N. G. Chu, L. N. Pfeiffer, and K. W. West. Tunable Fano interference in intersubband absorption. Optics Letters 1996 Jul 1;21(13):985-987.
72. J. Faist, F. Capasso, C. Sirtori, K. W. West, and L. N. Pfeiffer. Controlling the sign of quantum interference by tunnelling from quantum wells. Nature (London) 1997 Dec 11 ; 390(6660):589-591.
73. G. P. Agrawal. Nonlinear Fiber Optics, 3th Ed. Boston : Academic Press;2001.
74. S. E. Harris and L. V. Hau. Nonlinear Optics at Low Light Levels. Physical Review Letters 1999 Jun 7;82(23):4611-4614.
75. E. Paspalakis, M. Tsaousidou, and A. F. Terzis. Coherent manipulation of a strongly driven semiconductor quantum well. Physical Revew B 2006 Mar 29;73(12):125344(5).
76. K. S. Yee. Numerical Solution of Initial Boundary Value Problems Involving Maxwell's Equations in Isotropic Media. IEEE Transactions on Antennas and Propagation 1966 May; AP-14(3):302-307.
77. G. Mur. Absorbing boundary conditions for the finite-difference approximation of the time-domain electromagnetic-field equations. IEEE Trans Electromagn Compat 1981 Nov;23(4):377-382.
78. A. Taflove and M. E. Brodwin. Numerical solution of steady-state electromagnetic scattering problems using the time-dependent Maxwell's equations. IEEE Transactions on Microwave

Theory and Technology 1975 Aug; 23(8):623-630.

Chapter 4. Quantum correlations in four-wave mixing and quantum- beat lasers

Xiangming Hu[1), Guangling Cheng[1,2), Jinhua Zou[1,3), Xiaoxia Li[1)
1) Huazhong Normal University
2) East China Jiaotong University
3) Yangtze University

Abstract: We present a collection of our recent schemes that have been proposed for continuous variable entanglement. Included in this collection are two types of optical systems. One is the four-wave mixing in resonantly or near-resonantly driven atomic systems. The numerical results and physical analyses are presented by using dressed atomic states and squeeze transformed cavity modes. Two dissipation channels are identified, through which the dressed atoms simultaneously absorb in the excitations from the pair of squeeze transformed modes. It is in the presence of such two channels that the entanglement is greatly enhanced and the best achievable state is the original Einstein-Podolsky-Rosen (EPR) entangled state. This scheme is applicable in the optical regime where atomic spontaneous emission has to be taken into account, unlike the two-step atomic reservoir engineering scheme, which is limited to the microwave regime. The other kind of systems is the quantum-beat laser with incoherent pump or coherent driving. Such a laser, when it operates well above threshold, produces entangled light. The numerical results and physical analyses are presented by using the collective modes of the lasing fields. The relative mode is decoupled from the active medium and thus remains in its vacuum state, while the sum mode operates well above threshold and displays sub-shot noise. The quantum-beat and the sum mode intensity noise reduction combine to yield entanglement between two bright beams and sub-Poissonian photon statistics of respective beams.

Introduction

Four-wave mixing [1-4] and quantum-beat lasers [5] are two different types of coherent optical phenomena, but they are closely related to each other. They not only yield the signals of new frequencies but also have the potentials for providing sources of non-classical light [6-17]. In such systems, the optical fields are in near-resonant interactions with the media and so the couplings of the generated fields to the media are much stronger than in parametric processes [18,19]. It is for the very reason that the quantum fluctuations of the media due to spontaneous emission are fed into the generated fields. Since the strong driving fields create the atomic coherence and induce the correlations of the cavity fields [4,5], the quantum fluctuations are reducible and the generated fields are pulled into non-classical states. Here we show that it is possible to obtain entangled light from four-wave mixing and the quantum-beat lasers.

Recently, a great number of atomic coherent effects have been revealed. Among others is coherent population trapping (CPT) [20] or electromagnetically induced transparency (EIT) [21-24]. The essence of this effect is that the atoms are driven, under certain conditions, into a coherent superposition state, which is called dark state. At the dark state the atoms no longer absorb light. It has been shown that, by using CPT/EIT, the four-wave mixing can be greatly enhanced or can be opened in the otherwise impossible regime [25-33]. It has also been predicted that four wave mixing is an alternative and efficient way to generate continuous variable entanglement [34-40], which is the key source in quantum information and quantum communication [41]. It becomes aware that for the four-wave mixing in the two-level system [36-40], the sum of variances of a pair of Einstein–Podolsky–Rosen (EPR) like operators is limited to 50% below the standard quantum limit. It is desirable to enhance entanglement. Although it is possible to establish a squeeze operator for the cavity fields and obtain EPR entangled light [7,34] when the applied and generated fields are detuned in the extreme case, one returns to the parametric case, where the interaction is so weak. One possible way is to use the atomic reservoir engineering, as has been proposed by Pielawa et al. [35]. By two-step operation, EPR entanglement can be obtained [42]. For the two step operations, different resonance frequencies of the two-level atoms and different initial coherent superposition of atomic states are required. It is clear that such requirements make the experiment complicated. Recently, it has been shown that entanglement is significantly enhanced through the auxiliary channel for dressed state population transfer [37,39].

One purpose of this chapter is to explore the coherent effects on the quantum correlations in resonant or near-resonant cases. Two different cases are considered. One involves the off-resonant but near-resonant interaction of a pair of coherent driving fields with the three-level atoms [38]. One of two driving fields and one of two generated fields are coupled to a common electronic dipole transition. The driving field and the generated field on the same transition have the frequency difference determined by the generalized Rabi frequency. In the other case, the three-level atoms in the Lambda configuration are also employed [40]. The driving fields are resonantly coupled to one dipole transition and the dipole forbidden transition, the signal fields are generated from the other dipole transition. Using this mechanism, one has the microwave or Raman-transition controlled EPR light entanglement, or the simultaneous generation of sum-frequency mixing and EPR light entanglement.

Quantum-beat lasers are a particular form of correlated spontaneous emission lasers (CEL's) [43-49]. Quantum-beat is formed by creating coherence between near degenerate atomic states, either excited states or ground states. In particular, a beam of three-level atoms in Vee configuration emit photons into two modes. The atomic upper levels are initially prepared in a coherent superposition or are coupled by a coherent field [13-17]. The fluctuations of the relative phase and the relative amplitude drop to the vacuum levels. In addition to this, as a different form, correlated spontaneous emission can be formed by creating coherence between a pair of states between which lasing transitions occur. One such example is a two-photon CEL [13-17] with a beam of three-level atoms in cascade configuration. The top and bottom states are initially prepared in a coherent superposition state. It was predicted that the phase noise is reduced by 50% below the vacuum noise level.

A quantum-beat laser can be initialized by using lasing without population inversion. The mechanism for lasing without inversion essentially is CPT/EIT: absorption is reduced while stimulated emission remains unaffected [50-56]. At the same time, the coherence leads to noise squeezing [57-61]. The fluctuations of the laser intensity are suppressed up to or more than 50% below the shot noise. We also note that noise squeezing is possible even when neither initial coherence nor external coherent driving is used. This is caused by dynamic noise reduction mechanism [62-65]. Combining the quantum-beat laser operation and the above quantum noise squeezing mechanisms, we predicted that the two lasing modes and the sum mode exhibit sub-shot noise [66-68]. Most recently, two-mode CEL's have been proposed to be used as an entanglement amplifier [69-77]. It was shown that the amplifier yields continuous variable entanglement for a short time. The entanglement tends to vanish as time goes on and the laser intensity increases. That means that the entanglement is existent only near above threshold, where the saturation has negligible influence. Well above threshold, the laser intensity has strong saturation effect and thus the quantum entanglement is destroyed.

It is desirable to devise an entanglement laser as an active device that operates well above threshold. As the parallel purpose of this chapter we will show that it is possible to generate entangled light using a quantum-beat scheme together with incoherent pump or coherent driving [78]. At the same time, entanglement between two modes is compatible with sub-Poissonian photon statistics of the respective modes and lasing operation well above threshold each other. This device produces two bright beams of entangled sub-Poissonian light. A generalization to the tripartite entanglement is also given [79].

Entanglement in four-wave mixing

We begin with recalling the entanglement conditions of the two cavity modes. A system is entangled if it is not separable. That is to say, the density operator ρ for the state cannot be expressed as a combination of the product state,

$$\rho \neq \sum_l q_l \rho_l^{(1)} \otimes \rho_l^{(2)}, \tag{1}$$

with $q_l \geq 0$ and $\sum_l q_l = 1$. For bipartite continuous variable entanglement between the two Gaussian state lasing modes $a_{1,2}$ we can use a sufficient criterion proposed in Refs. [80]. According to the criterion, the two lasing fields are entangled if the sum of the variances of two EPR-like operators u and v of the two modes satisfies the inequality

$$V = \langle(\delta u)^2\rangle + \langle(\delta v)^2\rangle < 2, \tag{2}$$

where

$$u = X_1 - X_2, \qquad v = P_1 + P_2. \tag{3}$$

Here the quadrature operators for the two modes a_l are defined as

$$X_l = \tfrac{1}{\sqrt{2}}(a_l e^{i\phi_l} + a_l^+ e^{-i\phi_l}), \quad P_l = \tfrac{-i}{\sqrt{2}}(a_l e^{-i\phi_l} - a_l^+ e^{i\phi_l}), \tag{4}$$

with ϕ_l being the phases of the cavity modes a_l, l=1,2.

In the mixing systems, as will be shown below, a pair of the cavity field operators can combine into another pair of two-mode squeeze-transformed operators by the relation $b_l = S^+(r) a_l S(r)$, where $S(r) = \exp(re^{-i\phi} a_1 a_2 - re^{i\phi} a_1^+ a_2^+)$. Explicitly, this transformation gives [81],

$$\begin{aligned} b_1 &= a_1 \cosh r - a_2^+ e^{i\phi} \sinh r, \\ b_2 &= a_2 \cosh r - a_1^+ e^{i\phi} \sinh r. \end{aligned} \tag{5}$$

At steady state one has $\langle b_l \rangle$=0, i.e.,

$$\langle a_1 \rangle \cosh r - \langle a_2^+ \rangle e^{i\phi} \sinh r = 0. \tag{6}$$

Defining $\langle a_l \rangle = r_l e^{-i\phi_l}$, l=1,2, we have the locked sum phase $\phi_1 + \phi_2 = -\phi$. Introducing the variables

$$X = \tfrac{1}{\sqrt{2}}(b_1 e^{i\phi_1} + b_1^+ e^{-i\phi_1} - b_2 e^{i\phi_2} - b_2^+ e^{-i\phi_2}), \tag{7}$$

$$P = \tfrac{-i}{\sqrt{2}}(b_1 e^{i\phi_1} - b_1^+ e^{-i\phi_1} + b_2 e^{i\phi_2} - b_2^+ e^{-i\phi_2}), \tag{8}$$

we have

$$u = e^{-r} X, \quad v = e^{-r} P, \tag{9}$$

Using these relations we rewrite the variance sum in the entanglement criterion as

$$V = e^{-2r}[\langle(\Delta X)^2\rangle + \langle(\Delta P)^2\rangle]. \tag{10}$$

It becomes aware that V=2 is the low bound for the separable states, while V=0 corresponds to the perfect EPR entanglement.

(1) EPR entanglement in off-resonantly driven mixing systems [38]. First we consider the Lambda system, as shown in Fig. 1. An ensemble of N atoms is placed in a two-mode optical cavity. Two metastable states of the μ-atom are denoted by $|1^\mu\rangle$ and $|2^\mu\rangle$ and the excited state is indicated by $|3^\mu\rangle$. Two external driving fields of circular frequencies ω_{d1} and ω_{d2} are applied to the dipole-allowed transitions $|1^\mu,2^\mu\rangle$–$|3^\mu\rangle$. Two cavity fields of circular frequencies ω_1 and ω_2 are generated from these two transitions, respectively. The master equation for the density operator ρ of the atom-field system is written in the dipole approximation and in an appropriate rotating frame as [82-84]

$$\dot{\rho} = -\frac{i}{\hbar}[H,\rho] + L\rho, \tag{11}$$

where the Hamiltonian takes the form

$$H = H_0 + H_I, \tag{12}$$

$$H_0 = \sum_{l=1,2}\sum_{\mu}^{N} \hbar[\Delta_l \sigma_{ll}^\mu + \Omega_l(\sigma_{3l}^\mu + \sigma_{l3}^\mu)], \tag{13}$$

$$H_I = \sum_{l=1,2}\sum_{\mu=1}^{N} \hbar g_l(a_l e^{-i\delta_l t}\sigma_{3l}^\mu + \sigma_{l3}^\mu a_l^+ e^{i\delta_l t}). \tag{14}$$

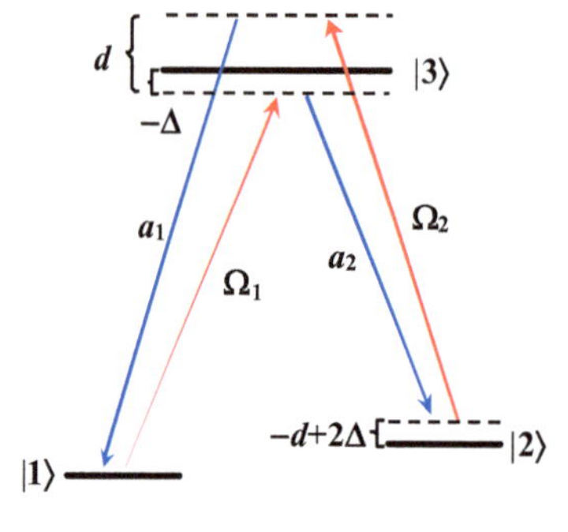

Figure 1: Three-level atoms with dipole transitions $|1,2\rangle$–$|3\rangle$ in the Lambda configuration. Two external coherent fields are coupled to the dipole transitions (half Rabi frequencies $\Omega_{1,2}$) and detuned asymmetrically with respect to respective transitions (Δ_1=$-\Delta_2$=Δ). Two cavity fields ($a_{1,2}$) are generated from Rabi sidebands.

Here $\sigma_{kl}^\mu = |k^\mu\rangle\langle l^\mu|$ (k,l=1,2,3) represents an atomic flip operator when $k \neq l$ and a projection operator when k=l for the μ-th atom. We assume that the phase factors due to the randomly orientated dipole moments of individual atoms are contained in the atomic states. a_l and a_l^+ are annihilation and creation operators for two cavity fields of frequencies ω_l, respectively; g_l are the coupling constants; Ω_l are half Rabi frequencies, Δ_l=$\omega_{dl}-\omega_{3l}$ and δ_l=$\omega_l-\omega_{dl}$ are detunings, l=1,2. The damping term $L\rho$ in Eq. (11) consists of two parts,

$$L\rho = L_a\rho + L_c\rho, \tag{15}$$

where $L_a\rho$ describes the atomic incoherent processes including the spontaneous decay and incoherent pump, and $L_c\rho$ represents for the cavity loss. For the present system $L_a\rho$ reads as

$$L_a\rho = \sum_{l=1,2} L_{l3}\rho, \tag{16}$$

where $L_{lk}\rho$ denotes the incoherent population transfer from state $|k\rangle$ to $|l\rangle$ at rate γ_{lk},

$$L_{lk}\rho = \sum_{\mu=1}^{N} \frac{\gamma_{lk}}{2}(2\sigma_{lk}^{\mu}\rho\sigma_{kl}^{\mu} - \sigma_{kl}^{\mu}\sigma_{lk}^{\mu}\rho - \rho\sigma_{kl}^{\mu}\sigma_{lk}^{\mu}). \tag{17}$$

The cavity loss of the two generated fields $L_c\rho$ takes the form

$$L_c\rho = \sum_{l=1,2} L_{a_l}\rho, \tag{18}$$

where $L_{a_l}\rho$ represents the loss of each mode a_l

$$L_{a_l}\rho = \sum_{l=1,2} \frac{\kappa_l}{2}(2a_l\rho a_l^{+} - a_l^{+}a_l\rho - \rho a_l^{+}a_l). \tag{19}$$

In what follows we will employ the dressed-atom squeeze-transformed mode approach to treat the present system.

(i) Dressed-atom representation. We assume that the Rabi frequencies associated with the driving field are much stronger than the atomic relaxation rates, $\Omega_k >> \gamma_{l3}$, $k,l=1,2$. For the sake of clarity, we consider the asymmetrical detunings $\Delta_1=-\Delta_2=\Delta$ and equal Rabi frequencies $\Omega_1=\Omega_2=\Omega$. The dressed states are obtained by diagonalizing Hamiltonian $H_0^{\mu} = \sum_{l=1,2} \hbar[\Delta_l \sigma_{ll}^{\mu} + \Omega_l(\sigma_{3l}^{\mu} + \sigma_{l3}^{\mu})]$ as [85]

$$\begin{aligned}
|+^{\mu}\rangle &= \tfrac{1}{2}(1+\sin\theta)\,|1^{\mu}\rangle + \tfrac{1}{2}(1-\sin\theta)\,|2^{\mu}\rangle + \tfrac{1}{\sqrt{2}}\cos\theta\,|3^{\mu}\rangle, \\
|0^{\mu}\rangle &= -\tfrac{1}{\sqrt{2}}\cos\theta\,|1^{\mu}\rangle + \tfrac{1}{\sqrt{2}}\cos\theta\,|2^{\mu}\rangle + \sin\theta\,|3^{\mu}\rangle, \\
|-^{\mu}\rangle &= \tfrac{1}{2}(1-\sin\theta)\,|1^{\mu}\rangle + \tfrac{1}{2}(1+\sin\theta)\,|2^{\mu}\rangle - \tfrac{1}{\sqrt{2}}\cos\theta\,|3^{\mu}\rangle,
\end{aligned} \tag{20}$$

where $\cos\theta = \sqrt{2}\Omega/d$, $\sin\theta = \Delta/d$, $d = \sqrt{\Delta^2 + 2\Omega^2}$. Using the dressed states, we rewrite the Hamiltonian H_0 as $\widetilde{H}_0 = \sum_{\mu=1}^{N} \hbar d(\sigma_{++}^{\mu} - \sigma_{--}^{\mu})$, which indicates that these dressed states $|0^{\mu}\rangle$ and $|\pm^{\mu}\rangle$ have eigenvalues $\lambda_{0,\pm}=0$, $\pm d$ and are equally spaced with the frequency gap d. It is easy to see that when $\Delta=0$, one has $|0^{\mu}\rangle = (-|1^{\mu}\rangle + |2^{\mu}\rangle)/\sqrt{2}$, which simply is the dark state, i.e., the coherent superposition of the metastable states [20]. This gives the maximal coherence of $-\frac{1}{2}$. However, no entanglement occurs in this case, as will be shown below. We will consider the off-resonant case of $\Delta \neq 0$.

(ii) Two-mode squeeze transformation. The master equation (11) can be rewritten in terms of the dressed atomic states as stated above and the squeeze transformed modes. We tune the cavity fields resonant with the Rabi sidebands $\delta_1=-\delta_2=d$. Now we make a unitary transformation with $\exp(-i\widetilde{H}_0 t/\hbar)$. After the transformation the interaction Hamiltonian and the density operator are denoted by the same symbols for simplicity. Since the dressed states are well separated from each other, $d >> \gamma_{13}$, γ_{23}, we can neglect rapidly oscillating terms such as $\exp(\pm idt)$ and $\exp(\pm 2idt)$, i.e., a rotating wave approximation is made [82-84]. Then the master equation (11) is rewritten in the form

$$\dot{\rho} = -\frac{i}{\hbar}[H_1, \rho] + L\rho, \tag{21}$$

where the Hamiltonian has the form

$$\begin{aligned}
H_1 = &-\sum_{\mu=1}^{N} \frac{\hbar g}{2}[\cos^2\theta a_1 - \sin\theta(1-\sin\theta)a_2^{+}]\sigma_{+0}^{\mu} \\
&-\sum_{\mu=1}^{N} \frac{\hbar g}{2}[\cos^2\theta a_2 - \sin\theta(1-\sin\theta)a_1^{+}]\sigma_{-0}^{\mu} + H.c.
\end{aligned} \tag{22}$$

In deriving we have assumed that $g_1=g_2=g$ for simplicity. The transitions described by Hamiltonian (22) are shown in Fig. 2(a). Substituting the squeezed transformed modes b_l in Eq. (5)

for a_l, we rewrite the Hamiltonian (22) as

$$H_{\mathrm{I}} = -\sum_{\mu=1}^{N} \hbar g_a (b_1 \sigma_{+0}^{\mu} + \sigma_{0+}^{\mu} b_1^{+} + b_2 \sigma_{-0}^{\mu} + \sigma_{0-}^{\mu} b_2^{+}) \quad \text{if } \tfrac{\Delta}{\Omega} > -\sqrt{\tfrac{2}{3}}, \tag{23}$$

$$H_{\mathrm{I}} = -\sum_{\mu=1}^{N} \hbar g_b (b_1 \sigma_{0-}^{\mu} + \sigma_{-0}^{\mu} b_1^{+} + b_2 \sigma_{0+}^{\mu} + \sigma_{+0}^{\mu} b_2^{+}) \quad \text{if } \tfrac{\Delta}{\Omega} < -\sqrt{\tfrac{2}{3}}, \tag{24}$$

where we have defined the coupling constants $g_a = \frac{1}{2} g(1-\sin\theta)\sqrt{1+2\sin\theta}$ for $\Delta > -\sqrt{\frac{2}{3}}\Omega$, $g_b = \frac{1}{2} g(1-\sin\theta)\sqrt{-1-2\sin\theta}$ for $\Delta < -\sqrt{\frac{2}{3}}\Omega$. The squeezing factor r for the transformation as in Eq. (5) determined by the classical field parameters,

$$r = \operatorname{arctan\,h} \frac{|\sin\theta|}{1+\sin\theta}, \qquad \Delta > -\sqrt{\tfrac{2}{3}}\Omega, \tag{25}$$

$$r = \operatorname{arctan\,h} \frac{1+\sin\theta}{|\sin\theta|}, \qquad \Delta < -\sqrt{\tfrac{2}{3}}\Omega, \tag{26}$$

and the phase ϕ is dependent on the sign of the detuning Δ,

$$\phi = \begin{cases} 0 & \text{for } \Delta > 0, \\ \pi & \text{for } \Delta < 0. \end{cases} \tag{27}$$

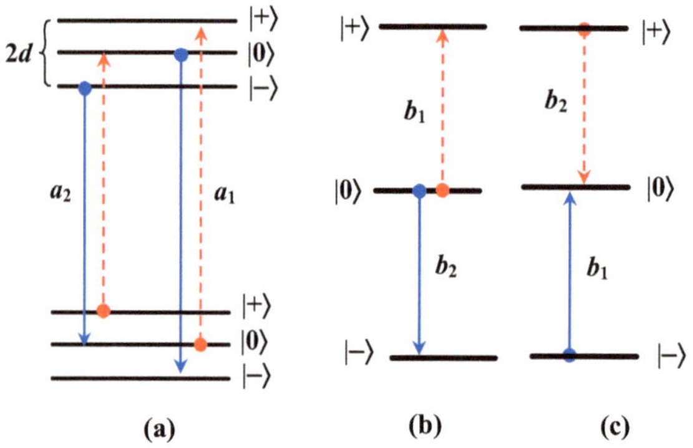

Figure 2: (a) The dressed transitions [Eq. (23)] in terms of the original cavity modes $a_{1,2}$. (b) The dressed transitions [Eq. (23)] in terms of squeeze-transformed modes $b_{1,2}$ for $\Delta > -\sqrt{\frac{2}{3}}\Omega$. (c) The dressed transitions [Eq. (24)] in terms of squeeze- transformed modes $b_{1,2}$ for $\Delta < -\sqrt{\frac{2}{3}}\Omega$.

The atoms behave differently depending on Δ/Ω. For $\Delta > -\sqrt{\frac{2}{3}}\Omega$, the atoms transit from $|0^{\mu}\rangle$ to $|\pm^{\mu}\rangle$ and absorb in the excitations from modes $b_{1,2}$, as shown in Fig. 2(b). The reversible processes, i.e., the transitions from $|\pm^{\mu}\rangle$ to $|0^{\mu}\rangle$ are responsible for the amplification of modes $b_{1,2}$, which are not shown. However, the former is always is dominant for the present case, since $\langle\sigma_{00}\rangle > \langle\sigma_{\pm\pm}\rangle$ is satisfied, as will be seen later. In sharp contrast, for $\Delta < -\sqrt{\frac{2}{3}}\Omega$, the atoms absorb in the excitations from modes $b_{2,1}$ and transit from $|\pm^{\mu}\rangle$ to $|0\rangle$, as shown in Fig. 2(c). The reversible processes are the amplification of modes $b_{2,1}$ and transitions from $|0^{\mu}\rangle$ to $|\pm^{\mu}\rangle$, which are not shown either. Similarly, the former is dominant due to $\langle\sigma_{\pm\pm}\rangle > \langle\sigma_{00}\rangle$. We note that the simultaneous presence of the Jaynes-Cummings (JC) and anti-JC interactions, appearing naturally in the trapped ions [86] but not in the context of cavity QED. As is well known, anti-JC terms in the context of cavity QED represent the nonconserving energy terms, corresponding to exciting (deexciting) the internal atomic state while creating (annihilating) an intracavity photon. Usually such terms are removed in the rotating wave approximation since their influence is negligibly weak. However, the case differs in the dressed-atom squeeze-transformed mode picture. The presence of anti-JC terms is determined by the fact that the strong driving fields provide energy. It has been shown that the simultaneous presence of JC and anti-JC is an important way for quantum information processing [87-89]. In the present case, JC and anti-JC couplings cause two-channel interaction of the dressed atoms with the squeeze-transformed modes. The above analysis shows that these two channels are responsible for the dissipation of the pair of squeeze-transformed modes. This is in sharp contrast to the two-level system, where only one such channel is present.

(iii) Damping term in terms of the dressed atomic states and the squeeze-transformed modes. Correspondingly, the damping term in Eq. (15) is rewritten as

$$L_a\rho = \sum_{l,k=+,-,0;l\neq k}(L_{lk}\rho + L_{lk}^{ph}\rho) + \sum_{l,k=+,-,;l\neq k} L_{lk}^{co}\rho, \tag{28}$$

$$L_{lk}^{ph}\rho = \varepsilon_{lk}\frac{\gamma_{lk}^{ph}}{4}(2\sigma_{lk}^{\mu p}\rho\sigma_{lk}^{\mu p} - \sigma_{lk}^{\mu p}\sigma_{lk}^{\mu p}\rho - \rho\sigma_{lk}^{\mu p}\sigma_{lk}^{\mu p}), \sigma_{lk}^{\mu p} = \sigma_{ll}^{\mu} - \sigma_{kk}^{\mu}, \tag{29}$$

$$L_{lk}^{co}\rho = \gamma_c(\sigma_{l0}^{\mu}\rho\sigma_{k0}^{\mu} + \sigma_{0k}^{\mu}\rho\sigma_{0l}^{\mu}), \tag{30}$$

$$\begin{aligned} L_c\rho = &\sum_{l=1,2}\frac{\kappa}{2}[N_0(b_l^+\rho b_l - b_l b_l^+\rho) + (N_0+1)(b_l\rho b_l^+ - b_l^+ b_l\rho)] \\ &+ \kappa M^*(b_1\rho b_2 + b_2\rho b_1 - b_1 b_2\rho - \rho b_1 b_2) + H.c, \end{aligned} \tag{31}$$

with ε_{lk}=1 for lk=0+, 0−, +−, otherwise ε_{lk}=0. We have assumed $\gamma_{13} = \gamma_{23} = \gamma$, $\kappa_1 = \kappa_2 = \kappa$. Other parameters in the above expressions are

$$\begin{aligned} &\gamma_{0+} = \gamma_{0-} = \tfrac{1}{2}\gamma\cos^4\theta, && \gamma_{+0} = \gamma_{-0} = \tfrac{1}{2}\gamma\sin^2\theta(1+\sin^2\theta), \\ &\gamma_{+-} = \gamma_{-+} = \tfrac{1}{4}\gamma\cos^2\theta(1+\sin^2\theta), && \gamma_{0+}^{ph} = \gamma_{0-}^{ph} = \tfrac{1}{4}\gamma\sin^2(2\theta), \\ &\gamma_{+-}^{ph} = \tfrac{1}{2}\gamma\cos^4\theta, && \gamma_c = \tfrac{1}{2}\gamma\sin^2\theta\cos^2\theta, \\ &N_0 = \sinh^2 r, && M = \sinh r\cosh r e^{i\phi}. \end{aligned} \tag{32}$$

In Eq. (28), $L_{lk}\rho$ has the same meaning as before except of the substitutions of the dressed states for the bare states, i.e., it describes the incoherent population transfer from one dressed state $|k^{\mu}\rangle$ to another $|l^{\mu}\rangle$ at rate γ_{lk} (l,k=0,±;$l\neq k$), $L_{lk}^{ph}\rho$ represents the phase damping between states $|l^{\mu}\rangle$ and $|k^{\mu}\rangle$, and $L_{lk}^{co}\rho$ stands for quantum coherence due to the interference between the incoherent processes. In the mixing case we can neglect the contribution of the cavity field to the populations [3,4]. Thus we obtain the steady state populations in the absence of the cavity fields as

$$\langle\sigma_{00}\rangle = \frac{\cos^4\theta}{1+3\sin^4\theta}, \quad \langle\sigma_{++}\rangle = \langle\sigma_{--}\rangle = \frac{1-\langle\sigma_{00}\rangle}{2}. \tag{33}$$

(iv) Master equation for the squeezed transformed modes. As usual, we can use the adiabatic elimination by assuming that the atomic variables decay much more rapidly than the cavity fields, γ>>κ. Adopting the standard linear laser theory [4,84], we can derive the master equation for reduced density operator ρ_c=Tr$_{\text{atom}}\rho$ of squeezed transformed cavity modes $b_{1,2}$. By tracing out the atomic variables, we obtain the master equation for the cavity fields from Eq. (21) as

$$\dot{\rho}_c = -\frac{i}{\hbar}\mathrm{Tr}_{atom}[H_1,\rho] + L_c\rho_c, \tag{34}$$

For the case of $\Delta > -\sqrt{\tfrac{2}{3}}\Omega$, the system dynamics is determined by the Hamiltonian (23) and correspondingly the master equation is derived as

$$\begin{aligned} \dot{\rho}_c = &\frac{1}{2}\sum_{l=1,2}\{(A+\kappa N_0)(b_l^+\rho_c b_l - b_l b_l^+\rho_c) + [B+\kappa(N_0+1)](b_l\rho_c b_l^+ - b_l^+ b_l\rho_c)\} \\ &+ (C+\kappa M^*)(b_1\rho_c b_2 + b_2\rho_c b_1) - (D_1+\kappa M^*)\rho_c b_1 b_2 - (D_2+\kappa M^*)b_1 b_2\rho_c \\ &+ H.c, \end{aligned} \tag{35}$$

where

$$A=\frac{2g_a^2N\Gamma\langle\sigma_{++}\rangle}{\Gamma^2-\gamma_c^2},\qquad B=\frac{2g_a^2N\Gamma\langle\sigma_{00}\rangle}{\Gamma^2-\gamma_c^2},\qquad C=\frac{g_a^2N\gamma_c(\langle\sigma_{++}\rangle+\langle\sigma_{00}\rangle)}{\Gamma^2-\gamma_c^2},$$
$$D_1=\frac{2g_a^2N\gamma_c\langle\sigma_{++}\rangle}{\Gamma^2-\gamma_c^2},\qquad D_2=\frac{2g_a^2N\gamma_c\langle\sigma_{00}\rangle}{\Gamma^2-\gamma_c^2},\tag{36}$$

with $\Gamma=\gamma_{0+}^{ph}+\frac{1}{2}(\gamma_{+-}+\gamma_{+0}+\gamma_{-0}+\gamma_{0+})+\frac{1}{4}(\gamma_{+-}^{ph}+\gamma_{0-}^{ph})$. A and B are the linear coefficients for gain and absorption, respectively, and C and $D_{1,2}$ describe the cross couplings between two modes, respectively. When absorption is dominant over gain, we have two dissipation for the two squeeze transformed modes, as will be shown below.

It is easy to derive the master equation for the case $\Delta<-\sqrt{\frac{2}{3}}\Omega$, in which the system dynamics is determined by the Hamiltonian (24). By changing the coupling constant g_a to g_b and exchanging the parameters $A\leftrightarrow B$, $D_1\leftrightarrow D_2$, we obtain the master equation for the case $\Delta<-\sqrt{\frac{2}{3}}\Omega$. This substitution is valid in the calculations of various correlations. In addition, we treat another case of cavity detunings $\delta_1=-\delta_2=-d$ in the same way.

(v) Calculation of the variances. Choosing the normal ordering [82-84] for the field operators and defining the c number variables $\beta_l\leftrightarrow b_l,\beta_l^*\leftrightarrow b_l^+$, we express the quantum fluctuations $\langle(\Delta X)^2\rangle+\langle(\Delta P)^2\rangle$ as

$$\langle(\Delta X)^2\rangle+\langle(\Delta P)^2\rangle=2+2[\langle\delta\beta_1^*\delta\beta_1\rangle+\langle\delta\beta_2^*\delta\beta_2\rangle$$
$$-\langle\delta\beta_1\delta\beta_2\rangle e^{i(\phi_1+\phi_2)}-\langle\delta\beta_1^*\delta\beta_2^*\rangle e^{-i(\phi_1+\phi_2)}]\tag{37}$$

in which various correlations can be calculated from the master equation (35). For the case of $\Delta>-\sqrt{\frac{2}{3}}\Omega$, the c-number Langevin equations for the cavity modes b_l are derived as

$$\dot\beta_1=\tilde A\beta_1-\tilde C\beta_2^*+F_{\beta_1},\tag{38}$$
$$\dot\beta_2=\tilde A\beta_2-\tilde C\beta_1^*+F_{\beta_2},\tag{39}$$

where the net gain coefficient $\tilde A$ and the cross coupling coefficients $\tilde C$ read

$$\tilde A=\frac{1}{2}(A-B-\kappa)=\frac{g_a^2N\Gamma(\langle\sigma_{++}\rangle-\langle\sigma_{00}\rangle)}{\Gamma^2-\gamma_c^2}-\frac{\kappa}{2},\tag{40}$$
$$\tilde C=D_1-C=\frac{g_a^2N\gamma_c(\langle\sigma_{++}\rangle-\langle\sigma_{00}\rangle)}{\Gamma^2-\gamma_c^2},\tag{41}$$

and the Langevin noise operators F_{β_1} and F_{β_2} have zero values $\langle F_{\beta_1}\rangle=\langle F_{\beta_2}\rangle=0$ and δ-correlations $\langle F_x(t)F_y(t')\rangle=2\langle D_{xy}\rangle\delta(t-t')$ with nonzero diffusion coefficients

$$2\langle D_{\beta_1^*\beta_1}\rangle=A+\kappa N_0,\qquad 2\langle D_{\beta_2^*\beta_2}\rangle=A+\kappa N_0,$$
$$2\langle D_{\beta_1\beta_2}\rangle=-(D_1+\kappa M),\qquad 2\langle D_{\beta_1^*\beta_2^*}\rangle=-(D_1+\kappa M^*).\tag{42}$$

The quantum correlations for the two modes b_1 and b_2 follow the dynamical equations

$$\frac{d}{dt}\langle\delta\beta_l^*\delta\beta_l\rangle=2\tilde A\langle\delta\beta_l^*\delta\beta_l\rangle-\tilde C(\langle\delta\beta_1\delta\beta_2\rangle+\langle\delta\beta_1^*\delta\beta_2^*\rangle)+2\langle D_{\beta_l^*\beta_l}\rangle,\quad l=1,2,\tag{43}$$
$$\frac{d}{dt}\langle\delta\beta_1\delta\beta_2\rangle=2\tilde A\langle\delta\beta_1\delta\beta_2\rangle-\tilde C(\langle\delta\beta_1^*\delta\beta_1\rangle+\langle\delta\beta_2^*\delta\beta_2\rangle)+2\langle D_{\beta_1\beta_2}\rangle.\tag{44}$$

In order to obtain the steady state solutions, we first consider the stability. Using Eqs. (5,38,39) and $\langle a_l\rangle=r_le^{-i\phi_l}$, l=1,2, we obtain a pair of equations for the evolution of the amplitudes of the modes a_l as

$$\dot r_1=\tilde Ar_1-\tilde C\cos(\phi_1+\phi_2)r_2,\tag{45}$$

$$\dot{r}_2 = \widetilde{A}r_2 - \widetilde{C}\cos(\phi_1 + \phi_2)r_1. \tag{46}$$

Using Eq. (27) we obtain the stability conditions as $\widetilde{A} \pm \widetilde{C} < 0$. In the good cavity limit, κ is negligibly small and then the stability conditions are simplified to $\widetilde{A} < 0$, i.e., $\langle\sigma_{00}\rangle > \langle\sigma_{\pm\pm}\rangle$. That means that the absorption is dominant over amplification. By setting time derivatives to zero, we obtain the steady state correlations from Eqs. (43,44)

$$\langle\delta\beta_1^*\delta\beta_1\rangle = \langle\delta\beta_2^*\delta\beta_2\rangle = -\frac{\widetilde{A}\langle D_{\beta_1^*\beta_1}\rangle + \widetilde{C}\langle D_{\beta_1\beta_2}\rangle}{\widetilde{A}^2 - \widetilde{C}^2}, \tag{47}$$

$$\langle\delta\beta_1\delta\beta_2\rangle = \langle\delta\beta_1^*\delta\beta_2^*\rangle = -\frac{\widetilde{A}\langle D_{\beta_1\beta_2}\rangle + \widetilde{C}\langle D_{\beta_1^*\beta_1}\rangle}{\widetilde{A}^2 - \widetilde{C}^2}. \tag{48}$$

The other quantum correlations at the steady state are calculated in the same way as

$$\langle(\delta\beta_1)^2\rangle = \langle(\delta\beta_2)^2\rangle = \langle\delta\beta_1^*\delta\beta_2\rangle = \langle\delta\beta_2^*\delta\beta_1\rangle = 0. \tag{49}$$

Making use of Eqs. (47-49), we finally obtain the variance sum of the EPR-like operators

$$V = 2e^{-2r}\left(1 - \frac{\langle D_{\beta_1^*\beta_1}\rangle + \langle D_{\beta_2^*\beta_2}\rangle \pm (\langle D_{\beta_1\beta_2}\rangle + \langle D_{\beta_1^*\beta_2^*}\rangle)}{\widetilde{A} \mp \widetilde{C}}\right). \tag{50}$$

In the numerator of Eq. (50), the "±" results from the factors $e^{\pm i(\phi_1+\phi_2)}$, in which the sum phase is shown in Eq. (27). We take "+" in the numerator and "–" in the denominator for Δ<0 and the other pair of signs are for Δ>0. By making the substitutions of the parameters as above, it is easy to obtain the results for the case $\Delta < -\sqrt{\frac{2}{3}}\Omega$. Especially, the stability condition $\langle\sigma_{00}\rangle < \langle\sigma_{\pm\pm}\rangle$ is obtained for such a case. The case for another kind of detunings $\delta_1 = -\delta_2 = -d$ is treated similarly.

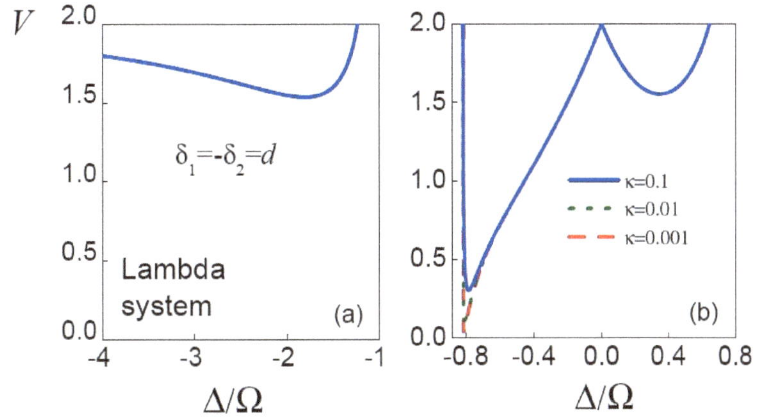

Figure 3: Variance sum V vs Δ/Ω for $\delta_1 = -\delta_2 = d$ in the Lambda system. The parameters are chosen as $g^2N=100$, κ =0.1 (solid); 0.01 (dotted); and 0.001 (dashed).

(vi) Results. In Fig. 3, we plot the variance sum V as a function of the ratio Δ/Ω for $\delta_1 = -\delta_2 = d$. We have scaled detunings, Rabi frequencies, coupling constants, and decay rates in units of γ. The parameters are chosen as $g^2N=100$ and κ=0.1 (solid), 0.01 (dotted) and 0.001 (dashed). We have: (i) $V \approx 0.304$ at Δ/Ω≈–0.78 for κ=0.1, (ii) $V \approx 0.112$ at Δ/Ω≈–0.80 for κ=0.01, and (ii) $V \approx 0.037$ at Δ/Ω≈–0.81 for κ=0.001. With the improvement of the quality factor of the cavity, the minimal value of the variance sum becomes smaller and smaller. For negligibly small rates of cavity loss, we have $V \approx 0$, which indicates that the best achievable state is the original EPR state. This is a pure state, $\mathrm{Tr}\rho_c = 1$. The above results show that the original EPR state is obtained when the driving fields are in near-resonant interaction with atoms and the cavity fields are in the resonant interaction with the dressed atoms via Rabi resonances (two channels). Physically, the two-channel interaction of the dressed atoms with the squeezed transformed modes is responsible for such quantum correlations. We note that for the two-level mixing system, there appears a single channel for the interaction of either of two combination modes with the dressed atoms, the variance sum is reduced at most by 50% below the upper bound of the inseparability (V=2). That is, the variance sum takes its minimal value V=1. However, for the present case, both squeeze transformed modes $b_{1,2}$ are mediated into the interaction with the dressed atoms through the two channels $|0^{\mu}\rangle \leftrightarrow |\pm^{\mu}\rangle$. As a result, the variance sum is doubly reduced, i.e., $V \rightarrow 0$, for a proper choice of parameters.

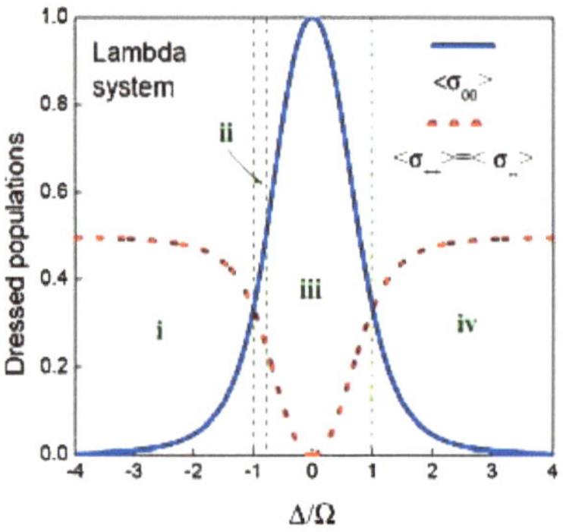

Figure 4: Dressed state populations $\langle\sigma_{00}\rangle$ (solid) and $\langle\sigma_{\pm\pm}\rangle$ (dotted) vs Δ/Ω for the same parameters as in Fig. 3.

Table I: Existence and inexistence of entanglement in different parameter regimes for $\delta_1 = -\delta_2 = d$.

Range of Δ/Ω	**$(-\infty,-1)$**	**$(-1,-\sqrt{2/3}$**	**$(-\sqrt{2/3},0)(0,1)$**	**$(1,\infty)$**
Dressed populations	$\langle\sigma_{\pm\pm}\rangle > \langle\sigma_{00}\rangle$	$\langle\sigma_{00}\rangle > \langle\sigma_{\pm\pm}\rangle$	$\langle\sigma_{00}\rangle > \langle\sigma_{\pm\pm}\rangle$	$\langle\sigma_{\pm\pm}\rangle > \langle\sigma_{00}\rangle$
Dominant process	Absorption	Amplification	Absorption	Amplification
Possibility of entanglement	Possible	No	Possible	No

Generally, the system dynamics can differ completely depending on the range of parameter Δ/Ω. The net gain coefficients $\tilde{A}$ and the cross-coupling coefficients $\tilde{C}$ for modes b_l depend on the dressed populations by the relations

$$\tilde{A} = \frac{g_a^2 N\Gamma(\langle\sigma_{++}\rangle - \langle\sigma_{00}\rangle)}{\Gamma^2 - \gamma_c^2} - \frac{\kappa}{2}, \quad \tilde{C} = \frac{g_a^2 N\gamma_c(\langle\sigma_{++}\rangle - \langle\sigma_{00}\rangle)}{\Gamma^2 - \gamma_c^2}, \quad \frac{\Delta}{\Omega} > -\sqrt{\frac{2}{3}}, \tag{51}$$

$$\tilde{A} = \frac{g_a^2 N\Gamma(\langle\sigma_{00}\rangle - \langle\sigma_{++}\rangle)}{\Gamma^2 - \gamma_c^2} - \frac{\kappa}{2}, \quad \tilde{C} = \frac{g_a^2 N\gamma_c(\langle\sigma_{00}\rangle - \langle\sigma_{++}\rangle)}{\Gamma^2 - \gamma_c^2}, \quad \frac{\Delta}{\Omega} < -\sqrt{\frac{2}{3}}. \tag{52}$$

Both the net gain and cross coefficients are relevant to the dressed population difference. Note that $\Gamma > \gamma_c$. Thus the net gain $\tilde{A}$ is proportional to $\langle\sigma_{++}\rangle - \langle\sigma_{00}\rangle$ for $\Delta > -\sqrt{\frac{2}{3}}\Omega$, but to $\langle\sigma_{00}\rangle - \langle\sigma_{++}\rangle$ for $\Delta < -\sqrt{\frac{2}{3}}\Omega$. The dressed population difference has a different sign or vanishes in a different range of parameter. When Δ=0, we have $\langle\sigma_{00}\rangle$=1 and $\langle\sigma_{\pm\pm}\rangle$=0. This is the very case where coherent population trapping occurs [20]. When $\Delta = \pm\Omega$, the population is equally distributed in three dressed states, $\langle\sigma_{00}\rangle = \langle\sigma_{\pm\pm}\rangle = 1/3$. For a general case, as shown in Fig. 4, we have

$$\langle\sigma_{00}\rangle > \langle\sigma_{\pm\pm}\rangle \quad \text{for } 0 < \tfrac{|\Delta|}{\Omega} < 1, \tag{53}$$

$$\langle\sigma_{00}\rangle < \langle\sigma_{\pm\pm}\rangle \quad \text{for } \tfrac{|\Delta|}{\Omega} > 1. \tag{54}$$

We can identify four different regimes for population distribution and the mode dissipation as follows. For clearness we list these cases in Tables I. (i) For $\frac{\Delta}{\Omega} < -1$, the system evolves according to Hamiltonian (23). Absorption in excitations from modes b_1 and b_2 occurs with the transitions from state $|\pm^{\mu}\rangle$, and the dressed population in state $|0^{\mu}\rangle$ causes amplification. However, the absorption is dominant over the amplification ($\widetilde{A} < 0$) since $\langle\sigma_{\pm\pm}\rangle > \langle\sigma_{00}\rangle$. In this case entanglement is existent, as shown in Fig. 3(a). (ii) For $-1 < \frac{\Delta}{\Omega} < -\sqrt{\frac{2}{3}}$, the system evolves following the same Hamiltonian as in case (i). However, we have different sign for the population difference, $\langle\sigma_{00}\rangle > \langle\sigma_{\pm\pm}\rangle$, i.e., $\widetilde{A} > 0$, which determines that amplification is more than absorption. In this case, the stability conditions are no longer satisfied and thus no steady state entanglement exists. (iii) For $-\sqrt{\frac{2}{3}} < \frac{\Delta}{\Omega} < 0$ and $0 < \frac{\Delta}{\Omega} < 1$, the system evolution is determined by Hamiltonian (24). The dressed population in state $|0^{\mu}\rangle$ gives absorption in excitations from modes b_1 and b_2 while the dressed population in state $|\pm^{\mu}\rangle$ leads to amplification. Since $\langle\sigma_{00}\rangle > \langle\sigma_{\pm\pm}\rangle$, absorption is dominant over amplification ($\widetilde{A} < 0$). Entanglement is present, as shown in Fig. 3(b). (iv) When $\frac{\Delta}{\Omega} > 1$, the system dynamics is controlled by the same Hamiltonian as in case (iii). In this case we have $\langle\sigma_{\pm\pm}\rangle > \langle\sigma_{00}\rangle$, Amplification dominates over absorption ($\widetilde{A} > 0$). However, the stability conditions are no longer satisfied and thus entanglement disappears.

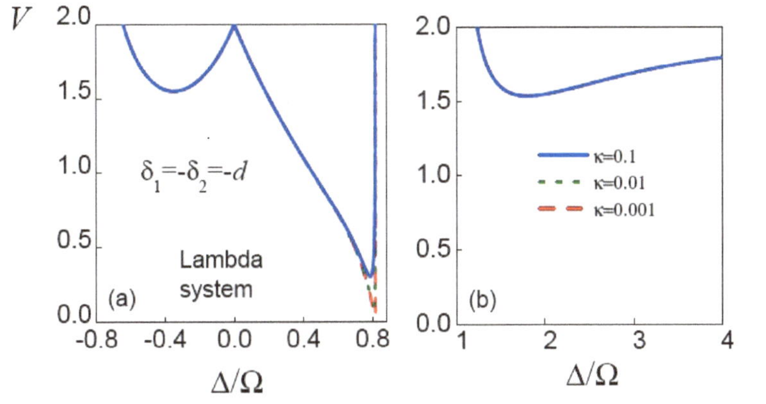

Figure 5: Variance sum V vs Δ/Ω for $\delta_1 = -\delta_2 = -d$ in the Lambda system. The parameters are the same as in Fig. 3.

Table II: Existence and inexistence of entanglement in different parameter regimes for $\delta_1 = -\delta_2 = -d$.

Range of Δ/Ω	**$(-\infty,-1)$**	**$(-1,0)(0,\sqrt{2/3})$**	**$(\sqrt{2/3},1)$**	**$(1,\infty)$**
Dressed populations	$\langle\sigma_{\pm\pm}\rangle > \langle\sigma_{00}\rangle$	$\langle\sigma_{00}\rangle > \langle\sigma_{\pm\pm}\rangle$	$\langle\sigma_{00}\rangle > \langle\sigma_{\pm\pm}\rangle$	$\langle\sigma_{\pm\pm}\rangle > \langle\sigma_{00}\rangle$
Dominant process	Amplification	Absorption	Amplification	Absorption
Possibility of entanglement	No	Possible	No	Possible

We turn to the case of $\delta_1 = -\delta_2 = -d$, in which the variance sum from the master equation can be obtained by following the same steps. In Fig. 5 we plot the variance sum V. In fact, curves in Figs. 5 and Fig. 3 are symmetrical with respect to $\Delta=0$. The minimal values are (i) $V \approx 0.304$ at $\Delta/\Omega \approx 0.78$ for $\kappa=0.1$ (solid), (ii) $V \approx 0.112$ at $\Delta/\Omega \approx 0.80$ for $\kappa=0.01$ (dotted), and (ii) $V \approx 0.037$ at $\Delta/\Omega \approx 0.81$ for $\kappa=0.001$ (dashed). Various cases are shown in Table II.

Next we present the case for three-level system in the Vee configuration as shown in Fig. 6. The master equation for the field-atom system has the same form as Eq. (11). Different from the Lambda system, the spontaneous decay in the Vee system occurs from two excited states $|1^{\mu},2^{\mu}\rangle$ to a single ground state $|3^{\mu}\rangle$, the damping term in the master equation reads as $L_a\rho = \sum_{l=1,2} L_{3l}\rho$.

The total Hamiltonian consists of two parts,

$$H_0 = \sum_{l=1,2}\sum_{\mu}^{N} \hbar[-\Delta_l \sigma_{ll}^{\mu} + \Omega_l(\sigma_{l3}^{\mu} + \sigma_{3l}^{\mu})], \tag{55}$$

$$H_I = \sum_{l=1,2}\sum_{\mu=1}^{N} \hbar g_l (a_l e^{-i\delta_l t}\sigma_{l3}^{\mu} + \sigma_{3l}^{\mu} a_l^{+} e^{i\delta_l t}), \tag{56}$$

where we have used $\Delta_l=\omega_{dl}-\omega_{l3}$, $\delta_l=\omega_l-\omega_{dl}$, l=1,2. Two coherent fields with the same Rabi frequencies $\Omega_1=\Omega_2=\Omega$ and asymmetrical detunings $\Delta_1=-\Delta_2=-\Delta$ lead to the same dressed states as for the Lambda system. Transforming into the rotating frame as in the previous case, taking the symmetrical parameters $\delta_1=-\delta_2=d$ and $g_1=g_2=g$, and using the squeeze transformation as shown in Eq. (5), we obtain the master equation of the same form as Eq. (21). The interaction Hamiltonian for the Vee system reads as

$$H_I = \sum_{\mu=1}^{N} \hbar g_a (b_1\sigma_{0-}^{\mu} + \sigma_{-0}^{\mu} b_1^{+} + b_2\sigma_{0+}^{\mu} + \sigma_{+0}^{\mu} b_2^{+}), \qquad \tfrac{\Delta}{\Omega} < \sqrt{\tfrac{2}{3}}, \tag{57}$$

$$H_I = \sum_{\mu=1}^{N} \hbar g_b (b_1\sigma_{+0}^{\mu} + \sigma_{0+}^{\mu} b_1^{+} + b_2\sigma_{-0}^{\mu} + \sigma_{0-}^{\mu} b_2^{+}), \qquad \tfrac{\Delta}{\Omega} > \sqrt{\tfrac{2}{3}}, \tag{58}$$

where $g_a = \frac{g}{2}(1+\sin\theta)\sqrt{1-2\sin\theta}$ and $g_b = \frac{g}{2}(1+\sin\theta)\sqrt{2\sin\theta-1}$. Here the squeezing factor r in the squeeze transformation is also determined by the classical field parameters, $r = \operatorname{arctanh}\frac{|\sin\theta|}{1-\sin\theta}$ for $\Delta < \sqrt{\frac{2}{3}}\Omega$, and $r = \operatorname{arctanh}\frac{1-\sin\theta}{|\sin\theta|}$ for $\Delta > \sqrt{\frac{2}{3}}\Omega$. At the same time, the phase ϕ depends on the sign of the detuning Δ. We have $\phi=\pi$ for $\Delta>0$, and $\phi=0$ for $\Delta<0$. There exist the simultaneous interactions of two squeeze-transformed modes $b_{1,2}$ with the dressed atoms, as shown in Hamiltonians (57) and (58). A slight difference is that the decay parameters become $\gamma_{\pm 0} = \frac{\gamma}{2}\cos^4\theta$, $\gamma_{0\pm} = \frac{\gamma}{2}\sin^2\theta(1+\sin^2\theta)$, while the other parameters remain the same as in Eq. (32). Following the same techniques, we obtain almost the same results as for the Lambda system. The variance sum is plotted in Figs. 7 and 8, respectively. The best achievable state approaches V=0. The three-level Vee system is as efficient as the three-level Lambda system for the enhancement of continuous variable entanglement. Also, it is the two-channel interaction that is responsible for the enhancement.

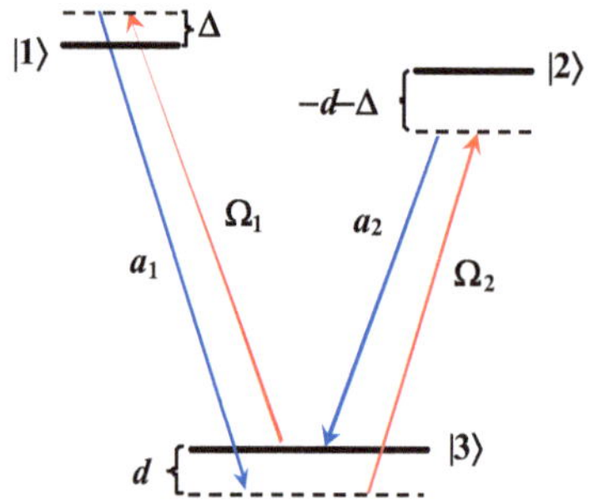

Figure 6: Three-level atoms with dipole transitions |1,2⟩−|3⟩ in the Vee configuration. Two external coherent fields are coupled to the dipole transitions (half Rabi frequencies $\Omega_{1,2}$) and detuned asymmetrically with respect to respective transitions ($\Delta_1 = -\Delta_2 = -\Delta$). Two cavity fields ($a_{1,2}$) are created via four-wave mixing.

Finally we can find advantages of the two-channel interaction over the one-channel interaction [35-37]. First, EPR entanglement ($V\approx 0$) is obtainable in the presence of two channels, while the variance sum is limited to $V\approx 1.0$ for the one-channel case. Second, the two-channel scheme is robust against spontaneous emission and is suitable for the entanglement generation in the optical frequency regime. However, the two-step reservoir engineering [35] based on the one channel interaction is established in the absence of spontaneous emission and so is limited to the microwave regime. Third, the present scheme does not need the two-step preparation the initial state, as required in the one channel scheme.

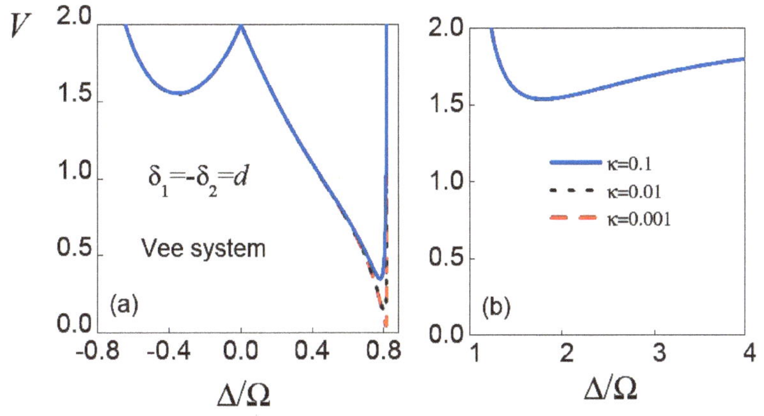

Figure 7: Variance sum V vs Δ/Ω for δ_1=$-\delta_2$=d in the Vee system. The parameters are the same as in Fig. 3.

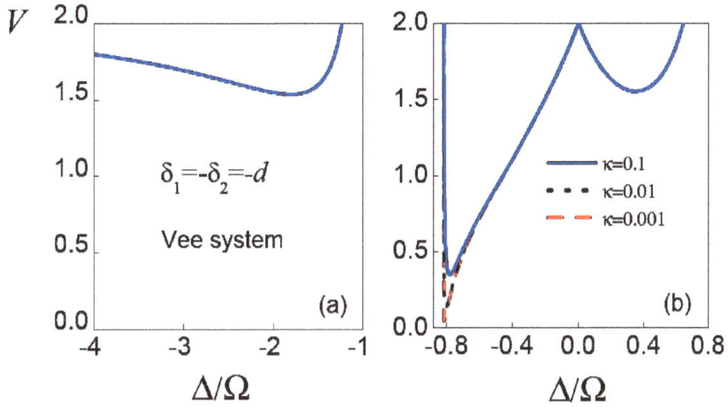

Figure 8: Variance sum V vs Δ/Ω for δ_1=$-\delta_2$=$-d$ in the Vee system. The parameters are the same as in Fig. 3.

Experimentally a gas of cold atoms can be used. A great number of atomic structures are suitable for the present schemes. For example, the hyperfine structures of the D_1 and D_2 lines of atomic ^{87}Rb are suitable for the experimental realization of the Lambda and Vee configurations. For the Lambda system we can use $|1\rangle = |5S_{1/2}, F=1\rangle$, $|2\rangle = |5S_{1/2}, F=2\rangle$, $|3\rangle = |5P_{3/2}, F=2\rangle$. Two hyperfine metastable states are separated by an amount of ω_{21}= 6.835 GHz. For the Vee system we can use $|1\rangle = |5P_{3/2}, F=3\rangle$, $|2\rangle = |5P_{1/2}, F=1\rangle$ and $|3\rangle = |5S_{1/2}\rangle$.

(2) EPR entanglement in resonantly driven mixing systems [40]. These systems appear in three different cases. First, the metastable states are coupled by a microwave field. It has turned out that microwave coupling is an efficient way for the optical coherent control. Such examples include four-wave mixing of optical and microwave fields [90], microwave-induced optical transparency [91,92], microwave controlled nonlinear optics [93], CEL's [43-46], lasers without population inversion [53]. So far the former four examples have been verified experimentally. This indicates that the microwave coupling does present a feasible mechanism for light control. As demonstrated recently [90], the fields differ in frequency by 5 orders of magnitude in the microwave mediated four wave mixing. Second, Metastable states are coupled to each other via a Raman transition. By substituting a Raman coupling for the microwave transition we obtain an equivalent coherent driving [94,95]. Third, metastable states are coupled via a two-photon transition. By such a mechanism, the sum-frequency generation (or even extreme ultraviolet radiation) is achieved by using EIT, as has been analyzed [96,97] and verified in recent 20 years [98–101]. An effective Rabi frequency can be used for the above three cases. This determines that the interaction Hamiltonians for these different schemes have the same form. Hence a unified model can be employed. The interaction configuration is essentially different from the previous one in that electronic dipole forbidden transition is employed and the driving fields and the cavity fields are

coupled to the different transitions.

In all three cases, an ensemble of N independent three-level atoms is placed in a two-mode optical cavity. The atomic dipole transitions $|1^{\mu},2^{\mu}\rangle-|3^{\mu}\rangle$ (or the spontaneous decay channels) are in the Lambda configuration, as shown in Fig. 9. $|1^{\mu}\rangle$ and $|2^{\mu}\rangle$ are metastable states, and $|3^{\mu}\rangle$ is the excited state. The metastable states are resonantly coupled to each other by a microwave field via magnetic dipole and/or electronic quadrupole transitions (a), or via a Raman transition (b), or via two-photon transition (c). Another coupling field couples the metastable state $|2^{\mu}\rangle$ to the excited state $|3^{\mu}\rangle$. Two cavity fields of circular frequencies ω_1 and ω_2 are generated on a different dipole allowed transition $|1^{\mu}\rangle-|3^{\mu}\rangle$. For the microwave coupling we assume that the level spacing between the metastable states (ω_{21}) is far larger than the Rabi frequency (Ω_0) associated with the microwave field $\omega_{21} >> \Omega_0$. This means that in all three cases the rotating wave approximation is valid. Since the cavity fields and the driving fields are coupled to the different transitions, the cavity fields have remarkably different frequencies from those of the applied optical fields. In particular, the two-photon excitation scheme yields the sum-frequency generation. The master equation for the density operator ρ of the atom–field system is written in the interaction picture takes the same from as Eq. (11). The system Hamiltonian reads as H=H_0+$H_{\rm I}$ with

$$H_0 = \sum_{\mu=1}^{N} \hbar[\Omega_0(\sigma_{12}^{\mu} + \sigma_{21}^{\mu}) + \Omega_1(\sigma_{23}^{\mu} + \sigma_{32}^{\mu})], \tag{59}$$

$$H_{\rm I} = \sum_{l=1,2}\sum_{\mu=1}^{N} \hbar g_l (a_l e^{-i\delta_l t}\sigma_{31}^{\mu} + \sigma_{13}^{\mu} a_l^{+} e^{i\delta_l t}), \tag{60}$$

where H_0 describes the resonant interaction of the driving fields with atoms, and $H_{\rm I}$ denotes the interaction of the cavity fields with the atoms. Ω_0 is effective half Rabi frequency for microwave transition (a), or for Raman transition (b), or for two-photon excitation (c), and Ω_1 is half Rabi frequency for the coupling field on the $|2\rangle-|3\rangle$ transition. For the sake of simplicity we have assumed that $\Omega_{0,1}$ are real. The detunings of the cavity field frequencies ω_l from the atomic resonance frequency ω_{31} are defined as $\delta_l = \omega_l - \omega_{31}$. The damping terms takes the same form as in the previous Lambda system. In what follows we use the same approach as above to present our analysis.

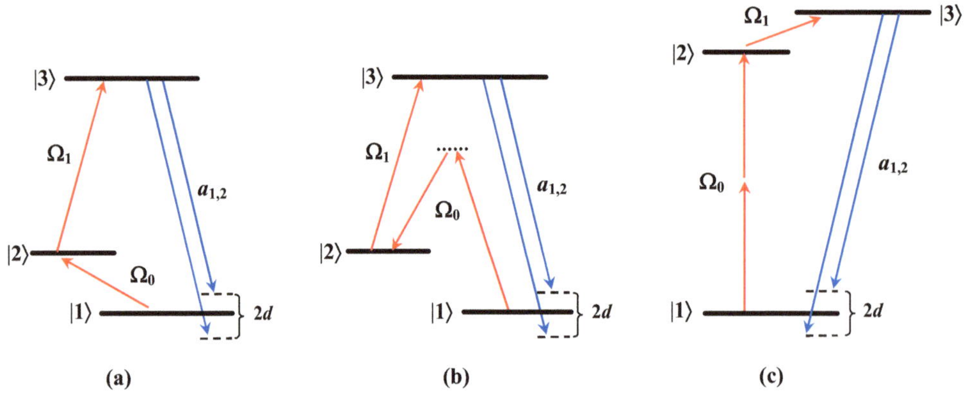

Figure 9: Three-level atoms with dipole transitions $|1,2\rangle-|3\rangle$ in the Lambda configuration. The dipole forbidden transition between two metastable states $|1\rangle$ and $|2\rangle$ are coupled to each other via microwave transition (a), or Raman transition (b), or two-photon transition (c) with effective half Rabi frequency Ω_0. Another optical field is resonantly coupled to the dipole transition $|2\rangle-|3\rangle$ with half Rabi frequency Ω_1. The two cavity fields $a_{1,2}$ are amplified from Rabi sidebands on the dipole transition $|1\rangle-|3\rangle$.

(i) Dressed-atom representation. Assuming that the driving fields are much stronger than atomic decay rates and the cavity modes and, $d = \sqrt{\Omega_1^2 + \Omega_0^2} >> \gamma_{l3}$, l=1,2, we obtain the dressed states by diagonalizing the Hamiltonian $H_0^{\mu} = \hbar\Omega_0(\sigma_{12}^{\mu} + \sigma_{21}^{\mu}) + \hbar\Omega_1(\sigma_{23}^{\mu} + \sigma_{32}^{\mu})$ for the μ-th atom and the driving fields as

$$|+^{\mu}\rangle = \tfrac{1}{\sqrt{2}}(\cos\theta\,|1^{\mu}\rangle + |2^{\mu}\rangle + \sin\theta\,|3^{\mu}\rangle),$$
$$|0^{\mu}\rangle = -\sin\theta\,|1^{\mu}\rangle + \cos\theta\,|3^{\mu}\rangle, \tag{61}$$
$$|-^{\mu}\rangle = \tfrac{1}{\sqrt{2}}(\cos\theta\,|1^{\mu}\rangle - |2^{\mu}\rangle + \sin\theta\,|3^{\mu}\rangle),$$

where $\tan\theta=\Omega_1/\Omega_0$. These dressed states $|0^{\mu}\rangle$ and $|\pm^{\mu}\rangle$ have their eigenvalues $\lambda_{0,\pm}=0,\pm d$ and are equally spaced with the frequency spacing d. In the dressed picture, the Hamiltonian H_0 is rewritten as $\widetilde{H}_0 = \sum_{\mu=1}^{N} \hbar d(\sigma_{++}^{\mu} - \sigma_{--}^{\mu})$. Correspondingly, the Hamiltonian H_I can be easily rewritten.

(ii) Two-mode squeeze transformation. The cavity fields are tuned resonant with the inner Rabi sidebands $\delta_1=-\delta_2=d$. By making the unitary transformation with $\exp(-i\widetilde{H}_0 t/\hbar)$ and neglecting the fast oscillating terms we rewrite the Hamiltonian H_I in the dressed-atom squeezed-transformed-mode picture as

$$H_I = \sum_{\mu=1}^{N} \hbar g_a(\sigma_{0-}^{\mu} b_1 + b_1^{+}\sigma_{-0}^{\mu} + \sigma_{0+}^{\mu} b_2 + b_2^{+}\sigma_{+0}^{\mu}), \qquad \frac{\Omega_1}{\Omega_0} < 1, \tag{62}$$

$$H_I = \sum_{\mu=1}^{N} \hbar g_b(\sigma_{+0}^{\mu} b_1 + b_1^{+}\sigma_{0+}^{\mu} + \sigma_{-0}^{\mu} b_2 + b_2^{+}\sigma_{0-}^{\mu}), \qquad \frac{\Omega_1}{\Omega_0} > 1, \tag{63}$$

where we have assumed $g_1=g_2=g$ and have defined $g_a = g\sqrt{\tfrac{1}{2}\cos(2\theta)}$ for $\Omega_1<\Omega_0$ and $g_b = -g\sqrt{-\tfrac{1}{2}\cos(2\theta)}$ for $\Omega_1>\Omega_0$. The squeezing factor r for the substitutions of the modes b_l for the modes a_l is determined by the classical field amplitudes. For $\Omega_1<\Omega_0$, we have $r=\text{arctanh}(\tan^2\theta)$, which increases with the increase of θ. When $\Omega_1>\Omega_0$, we have $r=\text{arctanh}(\tan^{-2}\theta)$, which decreases with the increase of θ. It is seen from Eqs. (62,63) that that the two-channel interaction (as shown in Fig. 2) of the squeezed transformed modes with dressed atoms are established. The dynamical behaviour of the dressed atoms is also strongly dependent on Ω_1/Ω_0. For $\Omega_1/\Omega_0<1$, the atoms absorb in the excitations from the modes $b_{2,1}$ and transit from $|\pm^{\mu}\rangle$ to $|0^{\mu}\rangle$, which corresponds to the case in Fig. 2(c). In sharp contrast, for $\Omega_1/\Omega_0>1$, the atoms absorb in the excitations from the modes $b_{1,2}$ and transit from $|0^{\mu}\rangle$ to $|\pm^{\mu}\rangle$, which corresponds to the case in Fig. 2(b). As in the previous case, the JC and anti-JC interactions are simultaneously realized.

By tracing out the atomic variables, i.e., $\rho_c=\text{Tr}_{\text{atom}}\rho$ we can obtain the information of the cavity fields. Following the standard techniques for the mixing interactions [4,84] we treat the cavity fields (b_1, b_2) linearly. Assuming that the atoms decay much more rapidly than the cavity fields, we can eliminate atomic variables adiabatically and derive the master equation of cavity modes $b_{1,2}$. For the case of $\Omega_1/\Omega_0<1$, where the Hamiltonian is shown in Eq. (62), the master equation for ρ_c has the same form as Eq. (35), and the parameters are the same as in Eq. (36). Also we have taken $\gamma_{13}=\gamma_{23}=\gamma$ and $\kappa_1=\kappa_2=\kappa$ for simplicity. The only difference lies in the dressed populations at steady state in the absence of the cavity fields. They are calculated as

$$\langle\sigma_{++}\rangle = \langle\sigma_{--}\rangle = \frac{\cos^2\theta + \cos^4\theta}{1+3\cos^4\theta} \quad \text{and} \quad \langle\sigma_{00}\rangle = \frac{\sin^4\theta}{1+3\cos^4\theta}, \tag{64}$$

which is dependent on the ratio Ω_1/Ω_0, not the detuning as in the previous case. We have $\langle\sigma_{\pm\pm}\rangle>\langle\sigma_{00}\rangle$ for $\Omega_1/\Omega_0 < \sqrt{2}$, and $\langle\sigma_{\pm\pm}\rangle<\langle\sigma_{00}\rangle$ for $\Omega_1/\Omega_0 > \sqrt{2}$, as shown in the inset in Fig. 10.

In exactly the same way we have the stability conditions $\widetilde{A} \pm \widetilde{C} < 0$. Note that $\gamma_c<\Gamma$ is always satisfied. In the good cavity limit, κ is negligibly small and then the stability conditions are reduced to $\widetilde{A} < 0$, i.e., $\langle\sigma_{\pm\pm}\rangle>\langle\sigma_{00}\rangle$. It is seen from the inset in Fig. 10 that the stability condition is always satisfied for the case of $\Omega_1/\Omega_0<1$. The variance sum is obtained from the master equation as

$$V = 2e^{-2r}\left(1 - \frac{A + D_1 + \kappa M + \kappa N_0}{\widetilde{A} + \widetilde{C}}\right). \tag{65}$$

It is easy to obtain the master equation for the case $\Omega_1/\Omega_0>1$, in which the system dynamics is described by the Hamiltonian (63). By exchanging parameters $A\leftrightarrow B$, $D_1\leftrightarrow D_2$ and changing coupling constant g_a to g_b in A, B, C, D_l, we obtain the master equation for the case $\Omega_1/\Omega_0>1$. Correspondingly, the steady-state solutions are obtained. Especially, the stability condition $\langle\sigma_{00}\rangle > \langle\sigma_{++}\rangle$ $(\widetilde{A} < 0)$ is obtained for the case $\Omega_1>\Omega_0$ in good cavity limit. From the population figure we find that $\langle\sigma_{00}\rangle > \langle\sigma_{++}\rangle$ is satisfied only when $\Omega_1 > \sqrt{2}\Omega_0$. It is obvious that in the region of $\Omega_0 \le \Omega_1 \le \sqrt{2}\Omega_0$, the system is unstable and so no steady solutions are existent.

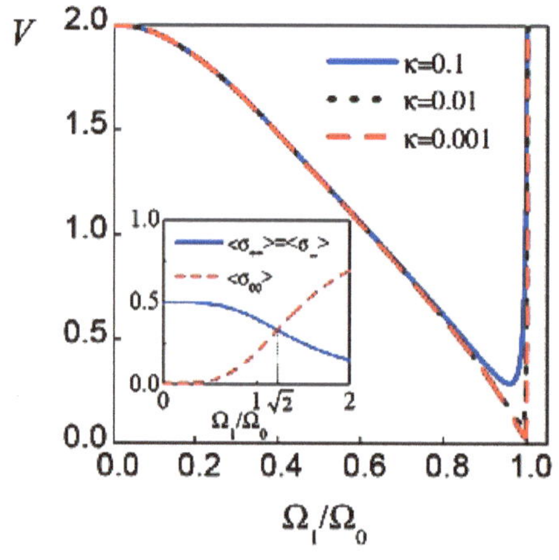

Figure 10: Variance sum V versus the ratio Ω_1/Ω_0. The parameters are chosen as $g^2N = 100$, $\kappa = 0.1$ (solid), 0.01 (dotted) and 0.001 (dashed). Plotted in the inset are the populations of dressed states $\langle\sigma_{\pm\pm}\rangle$ (solid) and $\langle\sigma_{00}\rangle$ (dotted).

Fig. 10 shows the variance sum V as a function of the ratio Ω_1/Ω_0. For the same scaling of parameters as above, we have chosen $g^2N=100$ and $\kappa=0.1$ (solid), 0.01 (dotted) and 0.001 (dashed). It is seen from Fig. 10 that the variance sum V first decreases gradually and then increases rapidly as the ratio Ω_1/Ω_0 increases. In the region of $0<\Omega_1/\Omega_0<1$, the value of V is always less than 2. This means that two cavity modes are entangled with each other at the steady state. For the different cavity loss rates, we have the minimal values: $V_{\min}\approx 0.285$ at $\Omega_1/\Omega_0=0.958$ ($\kappa=0.1$), $V_{\min}\approx 0.100$ at $\Omega_1/\Omega_0=0.987$ ($\kappa=0.01$), and $V_{\min}\approx 0.033$ at $\Omega_1/\Omega_0=0.996$ ($\kappa=0.001$). For the high-Q optical cavity, the variance sum $V\approx 0$ and thus the best achievable state approaches the original EPR state. Two factors are responsible for entanglement enhancement. The first is the two-channel interaction between the squeezed transformed modes and the dressed atoms. In the present system, each inner sideband mode simultaneously couples to two atomic dressed transitions and this leads to the creation of two dissipation channels, through which the atoms simultaneously absorb in the excitations from two combination modes and pull the cavity fields into an EPR entangled state. The second is the dependence of the squeezing factor r on the intensities of the external driving fields: $r = \operatorname{arctan h}(\Omega_1^2/\Omega_0^2)$ for $\Omega_1<\Omega_0$ and $r = \operatorname{arctan h}(\Omega_0^2/\Omega_1^2)$ for $\Omega_1>\Omega_0$. It is obvious that near at $\Omega_1/\Omega_0=1$, r is very large and then the variance sum is greatly reduced, $V\approx 0$, which corresponds to the perfect EPR entanglement.

Three differences exist between the two schemes stated above. (i) the EPR entangled state is obtained under the resonant driving conditions for the latter scheme, while for the former scheme it is impossible to obtain entanglement under resonance conditions. That is because complete destructive interference occurs on resonance and no mixing signals are existent in that scheme. (ii) Entanglement is accompanied with frequency conversion in the latter scheme but it is absent in the former scheme. In the former scheme, the applied and generated fields are coupled to the same transitions and so have almost the same frequencies. The latter scheme is suitable for the microwave coupling, the Raman transition driving and the two-photon exciting sum-frequency generation. In the microwave coupling case, a microwave field mediates into the coherent control of the optical field correlations. In the two-photon exciting scheme, optical sum-frequency generation is obtained. (iii) The squeezing factor for the latter system is determined by the ratio of amplitudes of applied fields Ω_1/Ω_0, while for the former scheme the squeezing factor is controlled by the ratio of the detunings to Rabi frequencies of the two applied coherent fields Δ/Ω. In the present system, EPR entanglement is obtained by varying the amplitudes of the applied fields. Near at $\Omega_1/\Omega_0=1$, the EPR state is prepared. In the former scheme, this is achieved by varying the frequencies of the driving fields.

Many atomic structures can be used for the above coupling schemes. For the microwave coupling scheme, for example, we can employ the atomic ^{87}Rb, $|1\rangle=|5S_{1/2},F=1\rangle$, $|2\rangle=|5S_{1/2},F=2\rangle$, $|3\rangle=|5P_{3/2},F=2\rangle$. Substituting a Raman transition for the microwave transition in the above scheme, one can realize the Raman scheme. Experimentally, the nonlinear frequency conversion was realized [94] by using atomic ^{208}Pb, $|1\rangle=|6s^2 6p^2\,^3P_0\rangle$, $|2\rangle=|6s^2 6p^2\,^3P_2\rangle$ and $|3\rangle=|6s^2 6p7s\,^3P_1\rangle$. In fact, when two beams of lasers at 293 nm and 425 nm are used to create the Raman coupling and a beam of laser at 460 nm couples the $|2\rangle$–$|3\rangle$ transition, a pair of sidebands centered around 283nm are amplified as the cavity fields. For the two-photon excitation sum-frequency mixing, a possible example is atomic hydrogen [98-101], where the 1s level is the ground state, the 2s level is the metastable state, and any of the *n*p (*n*=3–8) levels acts as the excited state. The 1s ground state and the 2s metastable state are coupled by a beam of laser at 243 nm, and the 3p–2s transition is driven by a strong coupling laser field at 656 nm. A pair of inner sidebands centered around the central wavelength 103 nm are generated from the Lyman (3p–1s) transition. The two-photon exciting field causes the photoionization and spoils the metastability of the state $|2\rangle$ to a certain degree. However, the photoionization can be suppressed by using coherent control techniques such as Stark chirped rapid adiabatic passage [102].

Entanglement in quantum-beat lasers

It is convenient to use a collective mode approach [47] to describe the quantum-beat lasers or CEL's. The collective modes are defined as

$$B_1 = \tfrac{1}{\sqrt{2}}(a_1 + a_2), \quad B_2 = \tfrac{1}{\sqrt{2}}(a_1 - a_2). \tag{66}$$

The Hermitian operators corresponding to the sum phase and sum amplitude are proportional to the real and imaginary parts of the sum mode operator B_1. Similarly, the real and imaginary parts of the relative mode operator B_2 determine the Hermitian operators corresponding the relative phase and relative amplitude [47]. Because of these correspondences, the mode B_1 is called the sum mode and the mode B_2 the relative mode. The biggest advantage of using the collective modes lies in the fact that, for CEL's, the relative mode B_2 is usually decoupled from the active media and simply undergoes absorption by the vacuum reservoir, while the sum mode B_1 is amplified and runs well above threshold, as will be shown below. This implies null amplitude for the relative mode B_2, $\langle B_2\rangle=0$. Consequently, we have $\langle a_1\rangle=\langle a_2\rangle$. This indicates that the phase difference of modes a_1 and a_2 is locked to $\phi_1-\phi_2=0$ [13-17].

Using the Langevin equations we obtain the photon number for the sum mode B_1 at steady state $\langle I_0\rangle = \langle B_1^+ B_1\rangle$. From now on we call it laser intensity without confusion. At the same time, the Mandel Q factor $Q = (\langle I_0^2\rangle - \langle I_0\rangle^2)/\langle I_0\rangle - 1$ can be calculated. The Mandel factor Q is the normally ordered normalized variance of the sum mode intensity and measures the deviations from Poissonian statistics. It is well-known that Q is zero for a coherent state (Poissonian photon statistics) and turns negative for a nonclassical state of an intracavity field (sub-Poissonian photon statistics). Using the Q factor, we obtain the normally ordered part of the output fluctuation spectrum [18,19]

$$S(\omega) = 2\int_0^\infty d\tau \cos(\omega\tau)\frac{\langle: i(t+\tau), i(t):\rangle}{\langle i(t)\rangle} = \frac{4Q\kappa\lambda}{\lambda^2+\omega^2}, \tag{67}$$

where $i(t) = \kappa\langle B_1^+(t)B_1(t)\rangle$ corresponds to the output photon flux operator and the inverse laser intensity correlation time λ is proportional to the differential gain. As is well known, $S(\omega)=0$ corresponds to shot noise and $-1\leq S(\omega)<0$ to sub-Poissonian photon statistics.

Using the variances in the sum and relative modes we obtain the Mandel factors $Q_{1,2}$ for the various modes $\langle I_l\rangle = \langle a_l^+ a_l\rangle = \frac{1}{2}\langle B_1^+ B_1\rangle$ as $Q_l = (\langle I_l^2\rangle - \langle I_l\rangle^2)/\langle I_l\rangle - 1 = \frac{1}{2}Q$, which indicates that sub-Poissonian statistics for the sum mode is compatible with sub-Poisssonian statistics for the respective modes. The amount of the noise reduction for the respective modes is half of that for the sum mode. Correspondingly, the normally ordered part of the output fluctuation spectra for the respective modes are expressed as

$$S_l(\omega) = \tfrac{1}{2}S(\omega). \tag{68}$$

Quadrature squeezing in the respective modes $a_{1,2}$ occurs when sub-Poissonian statistics in the sum mode is existent [$S(\omega)<0$].

The operators u and v are expressed in terms of the operators B_1 and B_2 as

$$u = B_1 e^{i\phi_1} + B_1^+ e^{-i\phi_1}, v = -i(B_2 e^{i\phi_1} - B_2^+ e^{-i\phi_1}), \tag{69}$$

which are the amplitude and phase quadratures corresponding to the sum mode B_1 and the relative mode B_2, respectively. Since the mode B_2 stays in its vacuum state, i.e., the variance in the quadrature operator v drops to its vacuum noise level [43-49], $\langle(\delta v)^2\rangle = 1$. For a laser far above threshold, the fluctuations in the amplitude are negligibly small compared with the amplitude itself. Under these circumstances, the variance of quadrature u is related to Mandel factor Q through the relation [103] $\langle(\delta u)^2\rangle = 1+Q$. Here 1 stands for the vacuum noise level. Using the above relations we obtain the output spectrum

$$V(\omega) = 2 + S(\omega), \tag{70}$$

which indicates that the two lasing fields are entangled if sub-Poissonian statistics in the sum mode is existent. It should be emphasized that the relation (70) is valid only for the present particular conditions, i.e., when the difference mode decouples and the sum mode operates sufficiently far above threshold.

Entanglement among more than two systems is going to be the key ingredient for multiparty communication in quantum teleportation networks [104], telecloning [105], and controlled dense coding [106]. In particular, entanglement for three modes has more classes than that for bipartite systems. The system density operator may be fully separable, or fully inseparable in any way, or partially separable. Correspondingly, classes of the tripartite entanglement range from one extreme, where the density matrix is not separable for any grouping of modes, to the other, where no entanglement appears between any two modes. A tripartite entangled state of continuous variables is an inseparable three-mode squeezed state that tends toward a Greenberger-Horne-Zeilinger [107] state in the limit of infinite squeezing. Optical nondegenerate parametric down conversion has turned out to be an effective way to prepare the tripartite continuous variable entanglement of optical fields [108,109]. Multipartite entangled beams are also produced by mixing squeezed vacuum with linear optical elements [110-112]. For the fully inseparable case, van Loock and Furusawa [113], who call this genuine tripartite entanglement, have derived a set of inequalities that are sufficient to detect it. Later Olsen and Bradley developed a single sufficient condition to detect genuine tripartite entanglement from the combined quadrature variances [114]. These inequalities read as

$$V_{ijk} = \langle(\Delta u_{ijk})^2\rangle + \langle(\Delta v_{ijk})^2\rangle) < 2, \tag{71}$$

with

$$u_{ijk} = X_i + \tfrac{1}{\sqrt{2}}(X_j + X_k), \quad v_{ijk} = P_i - \tfrac{1}{\sqrt{2}}(P_j + P_k), \tag{72}$$

where the mode indices i,j,k are all different. When any two of these inequalities are satisfied, the system is fully inseparable and genuine tripartite entanglement is guaranteed.

We define the collective modes [79]

$$B_{1,2} = \tfrac{1}{2}(a_1 + a_3) \pm \tfrac{1}{\sqrt{2}} a_2, \quad B_3 = \tfrac{1}{\sqrt{2}}(a_1 - a_3). \tag{73}$$

For convenience we call B_1 the sum mode and $B_{2,3}$ the difference modes. Then we rewrite the variance in inequalities [Eq. (71)] as

$$V_{213} = V_{231} = V(B_1 e^{i\phi_1} + B_1^+ e^{-i\phi_1}) + V(iB_2 e^{i\phi_1} - iB_2^+ e^{-i\phi_1}). \tag{74}$$

If the difference modes $B_{2,3}$ decouple, the phase differences between three modes are locked, $\phi_1-\phi_2=\phi_2-\phi_3=\phi_1-\phi_3=0$, i.e., $\phi_1=\phi_2=\phi_3$. At the same time the variances of quadrature drop to the vacuum level, $V(i(B_l - B_l^+)) = 1$, l=2,3. When the sum mode runs well above threshold, i.e., the steady state photon number is far larger than unity, the quadrature variance is related to the Mandel Q factor by the same relation as in the two-mode case [103], i.e., $\langle(\delta u)^2\rangle = 1+Q$. The intensities $\langle I_l \rangle = \langle a_l^+ a_l \rangle$ of the original cavity fields a_l (l=1,2,3) are related to the sum mode photon number $\langle I_0 \rangle = \langle B_1^+ B_1 \rangle$ by the relation $\langle I_1 \rangle = \langle I_3 \rangle = \tfrac{1}{4}\langle I_0 \rangle$, $\langle I_2 \rangle = \tfrac{1}{2}\langle I_0 \rangle$. Using the Q factors for the sum mode B_1 and the vacuum states for the decoupled modes $B_{2,3}$, we have the Mandel Q factors for the original cavity fields $Q_1 = Q_3 = \tfrac{1}{4}Q$, $Q_2 = \tfrac{1}{2}Q$. Correspondingly, we have the output fluctuation spectra for the respective modes

$$S_1(\omega) = S_3(\omega) = \tfrac{1}{4}S(\omega), \qquad S_2(\omega) = \tfrac{1}{2}S(\omega). \tag{75}$$

Then the correlation V_{213} is expressed as $V_{213}=V_{231}=2+Q$. Using the relation between Mandel Q factor and the output fluctuation spectrum $S(\omega)$, we obtain the quantum correlations for the output fields

$$V(\omega) = V_{213}(\omega) = V_{231}(\omega) = 2 + S(\omega). \tag{76}$$

If $S(\omega)<0$, corresponding to correlation $V_{213}(\omega)$ is lower than the quantum limit 2, it follows that the tripartite entanglement criterion is satisfied.

(1) A quantum-beat laser with dynamical noise reduction [66,78]. A pumping and coupling scheme is shown in in Fig. 11. An ensemble of N four-level atoms is placed in a two-mode cavity. The individual atoms are uncorrelated and the atomic dipole moments are randomly orientated. The atoms are pumped from the ground state $|0^\mu\rangle$ to the top state $|3^\mu\rangle$, emit photons into two lasing modes $a_{1,2}$ of circular frequencies $\omega_{1,2}$ through the $|1^\mu,2^\mu\rangle-|3^\mu\rangle$ transitions, and then return to the ground state. An external microwave field of circular frequency ω_v is resonantly coupled to the $|1^\mu\rangle-|2^\mu\rangle$ transition. As a consequence, two cavity modes are in the quantum-beat correlation. We will show that the above recycling of the atoms and the quantum-beat correlation combine to bring out the entanglement between two cavity fields.

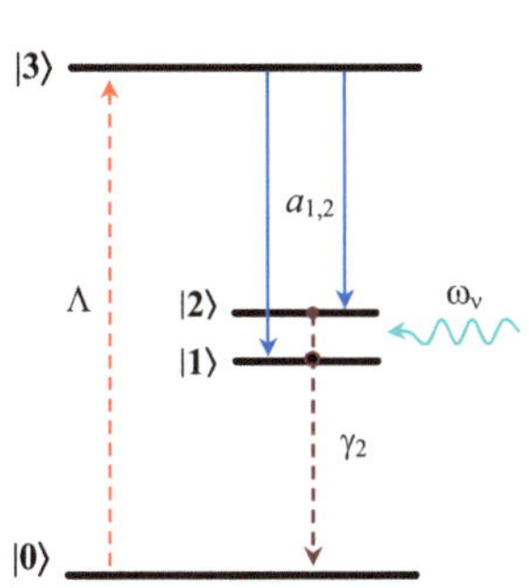

Figure 11: A quantum-beat laser scheme with dynamical noise reduction. Quantum-beat is created by using a microwave field to couple the two closely spaced states $|1\rangle$ and $|2\rangle$. Atoms are pumped incoherently from the ground state $|0\rangle$ to the top state $|3\rangle$ with rate Λ, emit photons into two lasing modes ($a_{1,2}$) on the $|1,2\rangle-|3\rangle$ transitions, and then return to the ground state at rate γ_2.

The master equation for the density operator of the system has the form as Eq. (11). The system Hamiltonian reads as $H=H_0+H_I$ with

$$H_0 = \sum_{\mu=1}^{N} \hbar\Omega_v (\sigma_{12}^{\mu} + \sigma_{21}^{\mu}), \tag{77}$$

$$H_I = \sum_{l=1,2}\sum_{j=1}^{N} \hbar g_l (a_l \sigma_{3l}^{\mu} e^{-i\delta_l t} + \sigma_{l3}^{\mu} a_l^{+} e^{i\delta_l t}), \tag{78}$$

where $\sigma_{kl}^{\mu} = |k^{\mu}\rangle\langle l^{\mu}|$ $(k,l=0,1,2,3)$ represents an atomic flip operator when $k\neq l$ and a projection operator when $k=l$ for the μ-th atom. We assume that the atomic states contain the phase factors due to the randomly orientated dipole moments. Ω_v is Rabi frequency associated with the microwave field and is assumed to be real. $\delta_l=\omega_l-\omega_{3l}$, $l=1,2$ are the detunings of the cavity fields from the corresponding atomic transitions. The atomic damping and pumping term reads

$$L_a\rho = \sum_{l=1,2} (L_{0l}\rho + L_{l3}\rho) + L_{30}\rho. \tag{79}$$

For simplicity we assume that $\gamma_{13}=\gamma_{23}=\gamma_1$, $\gamma_{01}=\gamma_{02}=\gamma_2$, and $\gamma_{31}=\Lambda$. The cavity loss term takes the same form as Eq. (19).

We assume that $\Omega_v>>(\gamma_l,\Lambda,|g_l\langle a_l\rangle|)$, $l=1,2$, and diagonalize Hamiltonian (77) and transform to the dressed picture. Then we obtain the dressed states as

$$|+^{\mu}\rangle = \tfrac{1}{\sqrt{2}}(|1^{\mu}\rangle + |2^{\mu}\rangle), \quad |-^{\mu}\rangle = \tfrac{1}{\sqrt{2}}(-|1^{\mu}\rangle + |2^{\mu}\rangle). \tag{80}$$

In terms of the dressed states we rewrite H_0 as $\widetilde{H}_0 = \sum_{\mu=1}^{N} \hbar\Omega_v(\sigma_{++}^{\mu} - \sigma_{--}^{\mu})$. Making a unitary transform with $\exp(-i\widetilde{H}_0 t/\hbar)$, tuning the cavity fields $\omega_l = \omega_{3l} - \Omega_v$, and neglecting the fast oscillating terms such as $\exp(\pm i\Omega_v t)$, we write the interaction Hamiltonian as

$$H_I = \sum_{\mu=1}^{N} \hbar g (B_1 \sigma_{3+}^{\mu} + \sigma_{+3}^{\mu} B_1^{+}), \tag{81}$$

where we have introduced the sum and difference modes B_1 and B_2 as in Eq. (66) and have assumed that $g_1=g_2=g$ for simplicity. We note that the difference mode B_2 is decoupled from the medium and only to the vacuum reservoir. This determines that the B_2 mode remains in the vacuum state. For the present system it is appropriate to assume that $\kappa_1=\kappa_2=\kappa$. The cavity loss term is rewritten in the form

$$L_c\rho = L_{B_1}\rho + L_{B_2}\rho. \tag{82}$$

Since the density operator for the B_2 mode can be separated from that for the composite system of the sum mode B_1 and the medium, we can write $\rho = \rho_{B_1}\rho_{B_2}$. This means that the master equation for the B_2 mode is reduced to

$$\dot{\rho}_{B_2} = L_{B_2}\rho_{B_2}, \tag{83}$$

Then we have the master equation for the B_1 mode and atoms

$$\dot{\rho}_{B_1} = -\frac{i}{\hbar}[H_I, \rho_{B_1}] + L\rho_{B_1}, \tag{84}$$

with the atomic and field damping term $L\rho_{B_1} = L_a\rho_{B_1} + L_{B_1}\rho_{B_1}$, where

$$L_a\rho_{B_1} = \sum_{l=+,-} (L_{l3}\rho_{B_1} + L_{0l}\rho_{B_1}) + L_{30}\rho_{B_1}, \tag{85}$$

and $L_{B_1}\rho_{B_1}$ has the same form as Eq. (19) except for the substitution of B_1 for a_l. By calculating the correlations of the collective modes $B_{1,2}$ and using the relation between the collective modes and the original modes $a_{1,2}$, we obtain the quantum statistical properties.

We derive Langevin equations from the master equation (84) by means of the generalized P representation of Drummond and Gardiner [115]. The atomic operators are defined as $\sigma_{kl} = \sum_{\mu=1}^{N} \sigma_{kl}^{\mu}$ $(k,l=0,\pm,3)$ and the correspondences between the c-numbers and the operators are set as $\beta\leftrightarrow B_1$, $\beta^*\leftrightarrow B_1^+$, $v\leftrightarrow\sigma_{+3}$, $v^*\leftrightarrow\sigma_{3+}$, $z_l\leftrightarrow\sigma_{ll}$. The set of Langevin equations are derived as

$$\dot{\beta} = -\tfrac{\kappa}{2}\beta - igNv + F_\beta, \tag{86}$$

$$\dot{v} = -(\gamma_1+\gamma_2)v_1 - ig\beta(z_3 - z_+) + F_v, \tag{87}$$

$$\dot{z}_0 = \gamma_2(z_+ + z_-) - \Lambda z_0 + F_{z_0}. \tag{88}$$

$$\dot{z}_3 = -\gamma_2 z_3 + \Lambda z_0 - ig\beta^* v + ig\beta v^* + F_{z_3}, \tag{89}$$

$$\dot{z}_- = -\gamma_1 z_3 + \gamma_2 z_{-0} + F_{z-}, \tag{90}$$

The populations follow the closure relation $\sigma_{00}+\sigma_{--}+\sigma_{++}+\sigma_{33}=1$. The noise correlations are easily obtained from Eqs. (86-90) by using the Einstein relations.

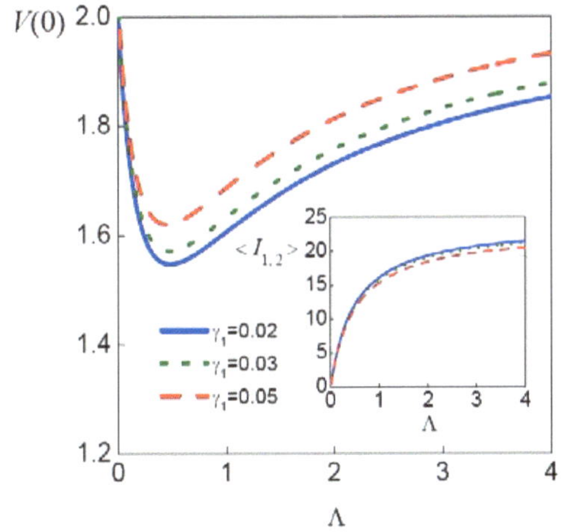

Figure 12: The zero-frequency output spectrum $V(0)$ as a function of for γ_1=0.02 (solid), 0.03 (dotted), and 0.05 (dashed) and C=200. The steady state intensities $\langle I_{1,2}\rangle$ in units of γ_2^2/g^2 are shown in the inset. All rates are in units of γ_2.

Plotted in Fig. 12 is the zero-frequency output spectrum $V(0)$ as a function of Λ. The incoherent pump rate Λ, the atomic decay rate γ_1 and the cavity loss rate κ (the cavity loss rates for the two modes are assumed to be equal to κ) are scaled in units of γ_2. The cooperativity parameter is defined as $C_0=2g^2N/(\kappa\gamma_2)$. The parameters are chosen as C_0=200, γ_1=0.02 (solid), 0.03 (dotted), 0.05(dashed). When large cooperativity parameter is large (C_0>>1), the system operates sufficiently far above threshold. The intensities $\langle I_l\rangle = \langle a_l^+ a_l\rangle$ (l=1,2) are plotted in the inset in units of γ_2^2/g^2. It is seen from Fig. 12 that the output spectrum $V(0)$ first decreases and then rises as the incoherent pump rates increases. At the same time, the photon number rises with the incoherent pump rate. On the other hand, as the decay rate γ_1 decreases the photon number increases and the output spectrum $V(0)$ decreases. In the limiting case ($\gamma_1\to 0$ and rate matching [62-65]), the spectrum $V(0)$ reaches its minimal value $\frac{3}{2}$. That shows that the entanglement criterion is satisfied when the quantum-beat laser works well above threshold. In addition, the output squeezing spectrum for the respective modes have their minimal value $S_l(0) = -\frac{1}{4}$, which corresponds to 25% noise reduction.

It can be easily deduced from the above that entanglement between two modes [$V(\omega)$<2] and squeezing in the respective modes are based on the combined effect of the correlated spontaneous emission ($\langle(\delta v)^2\rangle$ =1) and the sum mode intensity noise reduction [$S(\omega)$<0]. Two factors are responsible for entanglement and squeezing. The first is the correlated spontaneous emission. Only the sum mode B_1 is coupled to the medium while the relative mode B is decoupled from the system. The second is the dynamical noise reduction. For $\gamma_1 << \gamma_2, \Lambda$, the population in state $|-^\mu\rangle$ is negligible. Then the four-level system in Fig. 11 is reduced to a three-level system ($|0^\mu\rangle$, $|+^\mu\rangle$ and $|3^\mu\rangle$). In the reduced system, the succession of $\leftrightarrow|3\rangle$ the two-step incoherent process and the laser transitions $|+^{\mu}\rangle \xrightarrow{\gamma_2} |0^{\mu}\rangle \xrightarrow{\Lambda} |3^{\mu}\rangle \xrightarrow{A} |+^{\mu}\rangle$ recycles and regularizes the active laser electrons. The above two factors combine to cause squeezing and entanglement. For experimental realization, we can use alkali atoms for the present level scheme, $|0\rangle=|n\mathrm{S}1/2\rangle$, $|1\rangle=|n\mathrm{P}1/2\rangle$, $|2\rangle=|n\mathrm{P}3/2\rangle$, and $|3\rangle=|n\mathrm{D}3/2,5/2\rangle$. However, the unidirectional incoherent pumping is a stringent requirement. This perhaps may be achieved by using ultrafast pulses or coherently controlled adiabatic passage [116,117].

(2) A quantum-beat laser with external coherent driving [67,78]. Shown in Fig. 13 is a pumping and coupling scheme. The atoms are pumped from the ground state $|0^{\mu}\rangle$ to the excited states $|l^{\mu}\rangle$ (l=1,2) to provide necessary population for the laser gain. An external coherent field of circular frequency ω_0 is applied to the $|2^{\mu}\rangle$–$|3^{\mu}\rangle$ transitions to create atomic coherence, by which the system can operate without population inversion. Atoms emit photons into the laser modes a_l of circular frequencies $\omega_{1,2}$. In the dynamics, the atoms recycle through the successive channels $|0^{\mu}\rangle \xrightarrow{\Lambda} |l^{\mu}\rangle \xrightarrow{\gamma_1} |3^{\mu}\rangle \xrightarrow{\Omega} |l^{\mu}\rangle \xrightarrow{\gamma_2} |0^{\mu}\rangle$ (l=1,2). A microwave wave couples $|1^{\mu}\rangle$ and $|2^{\mu}\rangle$ and induces quantum-beat between the cavity fields. The recycling of the atoms and the quantum-beat pull the cavity fields into an entangled state.

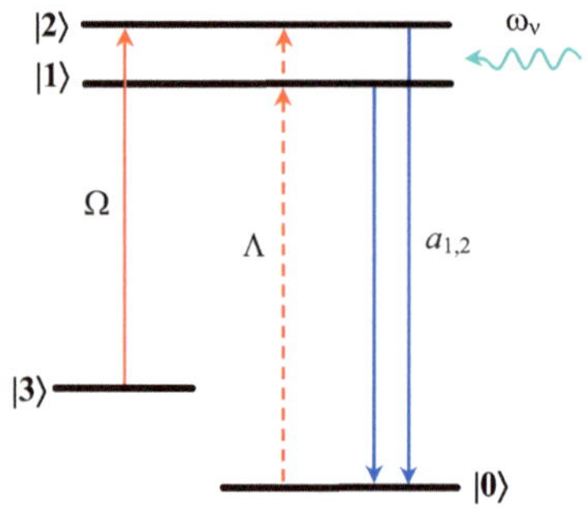

Figure 13: A quantum-beat laser scheme with coherent drive. Quantum-beat is created by using a microwave field to couple the two closely spaced states $|1\rangle$ and $|2\rangle$. Atoms are pumped incoherently from the ground state $|0\rangle$ to the excited states $|1, 2\rangle$ with rate Λ. Atomic coherence is created by coupling an external field to the $|2\rangle$–$|3\rangle$ with half Rabi frequency Ω. Atoms emit photons into two lasing modes ($a_{1,2}$) on the $|1,2\rangle$–$|3\rangle$ transitions.

The master equation for the density operator of the system is in the same form as Eq. (11). The system Hamiltonian takes the form $H=H_0+H_\mathrm{I}$ with

$$H_0=\sum_{\mu=1}^{N}\hbar\Omega_\mathrm{v}(\sigma_{12}^{\mu}+\sigma_{21}^{\mu}), \tag{91}$$

$$H_\mathrm{I}=\sum_{\mu=1}^{N}\hbar\Omega(e^{-i\Delta_0 t}\sigma_{23}^{\mu}+e^{i\Delta_0 t}\sigma_{32}^{\mu})+\sum_{l=1,2}\sum_{j=1}^{N}\hbar(g_l^{\mu}a_l e^{-i\delta_l t}\sigma_{l0}^{\mu}+g_l^{\mu *}a_l^{+}e^{i\delta_l t}\sigma_{0l}^{\mu}), \tag{92}$$

where $\Delta_0=\omega_0-\omega_{20}$, $\delta_l=\omega_l-\omega_{l0}$ (l=1,2) are detunings. Ω is Rabi frequency. The atomic damping term reads

$$L_a\rho=\sum_{l=1,2}(L_{0l}\rho+L_{l0}\rho+L_{3l}\rho), \tag{93}$$

We assume that $\gamma_{3l}=\gamma_1$, $\gamma_{l0}=\Lambda$, and $\gamma_{0l}=\gamma_2+\Lambda$, l=1,2, where $\gamma_{1,2}$ are spontaneous decay rates while Λ is incoherent pump rate. The cavity loss term is the same as in the previous scheme.

Following the same techniques we transform to the picture dressed by the microwave field and obtain the master equation as Eq. (84). The interaction Hamiltonian reads as

$$H_\mathrm{I}=\sum_{\mu=1}^{N}\frac{\hbar}{\sqrt{2}}\Omega(\sigma_{0+}^{\mu}+\sigma_{0+}^{\mu})+\sum_{\mu=1}^{N}\hbar g(B_1\sigma_{+0}^{\mu}+\sigma_{0+}^{\mu}B_1^{+}), \tag{94}$$

where we have tuned the coherent driving and the cavity fields as $\Omega_0=\Delta_l=\Omega_\mathrm{v}$, and have assumed that $g_1=g_2=g$. It is seen from Eq. (94) that the sum mode B_1 enters the interaction while the mode B_2 mode is decoupled. At the same time the damping term is written as $L\rho_{B_1} = L_a\rho_{B_1} + L_{B_1}\rho_{B_1}$, where

$$L_a\rho_{B_1} = \sum_{l=+,-}(L_{0l}\rho_{B_1} + L_{l0}\rho_{B_1} + L_{3l}\rho_{B_1}), \tag{95}$$

and $L_{B_1}\rho_{B_1}$ has the same form as Eq. (19) except for the substitution of the mode B_1 for a_l. Also we have assumed that $\kappa_1=\kappa_2=\kappa$ for simplicity.

Plotted in Fig. 14 is the zero-frequency output spectrum $V(0)$ as a function of the cooperativity parameter $C_0=2g^2N/(\kappa\gamma_2)$ (equal cavity loss rates for two cavity modes) for Ω =1.00 (solid), 1.25 (dotted), 1.50 (dashed), γ_1=4.0, and Λ =0.5. Such a choice of the parameters ($\Lambda<\gamma_2$) indicates that the system operates without population inversion [50-56]. We have scaled the parameters γ_1, Λ, Ω and κ in units of γ_2. The intensities $\langle I_1\rangle=\langle I_2\rangle$ in units of γ_2^2/g^2 are plotted in the inset. The characteristic features are shown as follows.

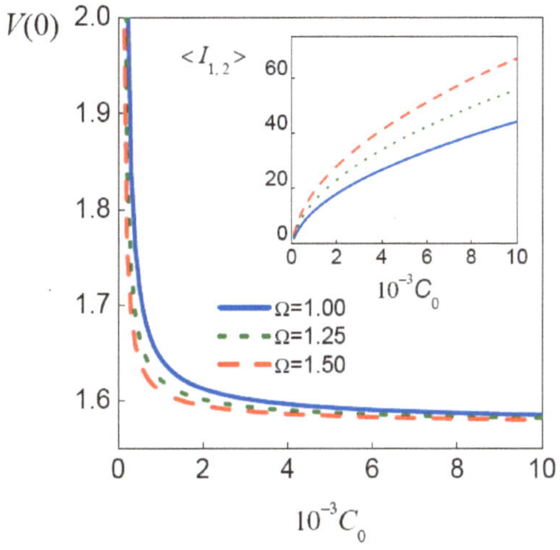

Figure 14: The zero-frequency output spectrum $V(0)$ versus the cooperativity parameter C_0 for Ω=1.00 (solid), 1.25 (dotted), and 1.50 (dashed) and γ_1=4.0, Λ=0.5. Shown in the inset are the steady state intensities $\langle I_{1,2}\rangle$ in units of γ_2^2/g^2. Rabi frequency and all rates are in units of γ_2.

(i) The laser runs well above threshold. The present system acts as an active device and the steady state intensities $\langle I_{1,2}\rangle$ for the modes $a_{1,2}$ are of order of tens of γ_2^2/g^2, as shown in the inset in Fig. 14. It is seen from the inset that the intensities increase as increasing cooperativity parameter. At the same time, the intensities is dependent on the other parameters. For example, appropriate increase in the driving field Rabi frequency leads to the increase in the laser intensities. Essentially, the two-mode laser is a laser without inversion [50-55]. The two modes are correlated and mediate into interaction in the form of a single sum mode B_1. We note that in the absence of incoherent pumping, the system is in the configuration of coherent population trapping or electromagnetically induced transparency [15-17]. However, it is the very essence of lasers without population inversion. In this case, absorption is cancelled due to atomic coherence. For the laser operation, transparency is spoiled by introducing incoherent processes, such as the incoherent pump. In fact, the population transfer, necessary for the light amplification, is achieved by using successive incoherent transitions $|0^{\mu}\rangle \xrightarrow{\Lambda} |\pm^{\mu}\rangle \xrightarrow{\gamma_1} |4^{\mu}\rangle$. Due to the combined effect of electromagnetically induced transparency and incoherent population transfer, the laser oscillation is established with no need of population inversion.

(ii) The two modes display sub-Poissonian statistics. It is seen from Fig. 14 that the output fluctuation spectrum $S(0)= V(0)-2$ for the sum mode B_1 takes its value $S(0)<0$ when the system operates well above threshold ($C_0 >> 1$). The spectrum $S(0)$ drops further downward from zero as C_0 rises. At the same time, the intensity fluctuations are controlled by the driving field Rabi frequency Ω. A large Rabi frequency leads to a small value in the spectrum $S(0)$. The best achievable squeezing is 50%, i.e., $S(0) \approx -\frac{1}{2}$. According to Eq. (68), the spectra for all modes are negative, $S_l(0)<0$. All of three original cavity fields have sub-Poissonian statistics. First, we note that the correlated spontaneous emission causes decoupling of the modes $B_{2,3}$. As a consequence, these two modes are in the vacuum states. Then, the responsible mechanism for the B_1 mode squeezing can be traced to the coherence controlled intrinsic feedback. On one hand, the atomic coherence ρ_{4+} or ρ_{40} plays its role in the laser gain and so the population inversion is not necessary. In the dressed-atom collective-mode picture, there are three channels for the population transfer from the lower lasing level $|0^\mu\rangle$ to the auxiliary level to $|4^\mu\rangle$: $|0^\mu\rangle \xrightarrow{\Lambda} |\pm^\mu, d^\mu\rangle \xrightarrow{\gamma_1} |4^\mu\rangle$. Then the electron is recycled through the coherent driving transition $|4^\mu\rangle \xrightarrow{\Omega} |+^\mu\rangle$ and the successive lasing transition $|+^\mu\rangle \xrightarrow{B_1} |0^\mu\rangle$. In the dynamics, the population in the excited states is small enough, and so the fluctuations from spontaneous emission become negligible. This pulls the mode B_1 to the vacuum noise level. On the other hand, the intrinsic feedback is important since the laser is operated well above threshold. The intensity diffusion coefficient turns negative when the laser intensity is large. This gives negative Mandel Q factor. In fact, the above factors combine to play their role. It is not difficult to image that in the absence of the driving field Ω, no light amplification occurs. That indicates that the noise squeezing of the mode B_1 is attributed to the coherence controlled intrinsic feedback. Finally, the squeezed mode B_1 and the vacuum modes $B_{2,3}$ correspond to the squeezing of all three original modes $a_{1,2,3}$. Therefore we conclude that quantum-beat and coherence controlled intrinsic feedback support the squeezing of all three laser modes.

(iii) Two fields are entangled. According to the criterion, one has two-mode entanglement when $V(0)<2$. As shown in Fig. 14, the entanglement criterion is satisfied for a wide range of parameters. In particular, the bipartite entanglement is obtained in the regime well above threshold. In other words, entanglement is achievable even when the saturation is deep. This is in sharp contrast to the case of the entanglement amplifier [69]. In that case, entanglement was predicted to be existent in a limited time period, but it disappears as the laser intensities rise. This indicates that entanglement only occurs near above threshold. The fine structures of the D_1 and D_2 line of atomic ^{87}Rb can be used for the present scheme $|0\rangle=|5S_{1/2},F=1\rangle$, $|1\rangle=|5P_{1/2}\rangle$, $|2\rangle=|5P_{3/2}\rangle$, and $|3\rangle=|5S_{1/2},F=2\rangle$

(3) Tripartite entangled light from a three-mode quantum-beat laser [79]. An ensemble of N atoms are placed in a three-mode optical cavity. The atoms have three excited states $|1^\mu\rangle$, $|2^\mu\rangle$ and $|3^\mu\rangle$ and two metastable states $|0^\mu\rangle$ and $|4^\mu\rangle$, as shown in Fig. 15. The upper levels may arise from Zeeman splitting or hyperfine structure. Laser fields of circular frequencies ω_l (l=1,2,3) are produced through the transitions from the excited states $|l^\mu\rangle$ to one of the metastable states $|0^\mu\rangle$. An external coherent field of circular frequency ω_0 is applied to the transition $|3^\mu\rangle$–$|4^\mu\rangle$. Two microwave fields of frequencies ω_{vl} (l=1,2) are coupled to the transitions $|1^\mu\rangle - |2^\mu\rangle$ and $|2^\mu\rangle - |3^\mu\rangle$, respectively. An incoherent field as an incoherent pump is applied to the transitions from the metastable state $|0^\mu\rangle$ to the excited states $|1^\mu,2^\mu,3^\mu\rangle$. In the present system, quantum beats between three cavity fields are created by coupling the microwave fields Ω_{vl} (l=1,2) to the transitions $|1^\mu\rangle \xrightarrow{\Omega_{v1}} |2^\mu\rangle \xrightarrow{\Omega_{v2}} |3^\mu\rangle$. Incoherent population pumping is performed through three-channel two-step pathways $|0^\mu\rangle \xrightarrow{\Lambda} |1^\mu,2^\mu,3^\mu\rangle$. The coherent transitions $|3^\mu\rangle \xrightarrow{\Omega} |4^\mu\rangle$ induces atomic coherence, which leads to lasing without population inversion [50-56]. Light amplification occurs on the correlated transitions: $|1^\mu,2^\mu,3^\mu\rangle \xrightarrow{a_{1,2,3}} |0^\mu\rangle$. As in two-mode quantum-beat laser with external coherent driving, the successive channels and the quantum-beats have the same effects on the correlations among three cavity fields.

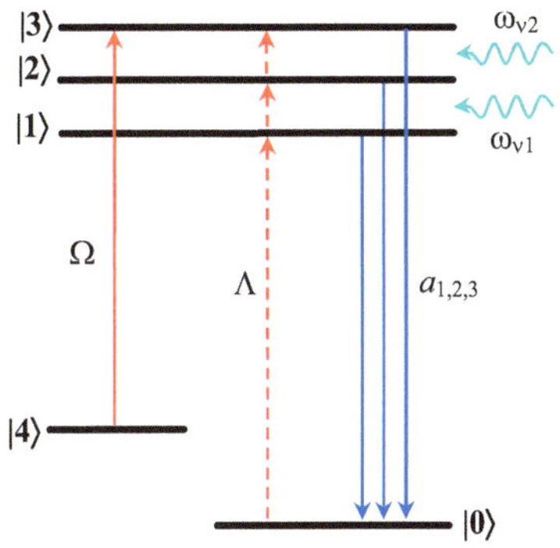

Figure 15: A three-mode quantum-beat laser scheme with coherent drive. Quantum-beat is created by using two microwave fields to couple the two closely spaced states $|1,2,3\rangle$. Atoms are pumped incoherently from the ground state $|0\rangle$ to the excited states $|1,2,3\rangle$ with rate Λ. Atomic coherence is created by coupling an external field to the $|3\rangle$–$|4\rangle$ with half Rabi frequency Ω. Atoms emit photons into three lasing modes ($a_{1,2,3}$) on the $|1,2,3\rangle - |0\rangle$ transitions.

The master equation for the density operator ρ of the atom-field system is written in the same form as Eq. (11). The system Hamiltonian reads as $H=H_0+H_I$ with

$$H_0 = \sum_{l=1}^{2}\sum_{\mu=1}^{N}\hbar\Omega_{vl}(\sigma^{\mu}_{l,l+1}+\sigma^{\mu}_{l+1,l}), \tag{96}$$

$$H_I = \sum_{\mu=1}^{N}\hbar\Omega(e^{-i\Delta_0 t}\sigma^{\mu}_{34}+e^{i\Delta_0 t}\sigma^{\mu}_{43})+\sum_{l=1}^{3}\sum_{\mu=1}^{N}\hbar g_l(a_l e^{-i\delta_l t}\sigma^{\mu}_{l0}+\sigma^{\mu}_{0l}a_l^{+}e^{i\delta_l t}), \tag{97}$$

where $\Delta_0=\omega_0-\omega_{34}$, $\delta_l=\omega_l-\omega_{l0}$ are detunings, l=1,2,3. Ω_{v1}, Ω_{v2} and Ω are half Rabi frequencies and are assumed to be real. The atomic damping term takes the form

$$L_a\rho = \sum_{l=1}^{3}(L_{0l}\rho + L_{l0}\rho + L_{4l}\rho), \tag{98}$$

where we assume that $\gamma_{4l}=\gamma_1$, $\gamma_{l0}=\Lambda$, and $\gamma_{0l}=\gamma_2+\Lambda$, l=1,2,3. The cavity loss term contains three parts

$$L_c\rho = \sum_{l=1}^{3}L_{a_l}\rho. \tag{99}$$

By using the dressed states and the collective modes, we rewrite the master equation in the form that is convenient for discussing steady state operations and quantum correlations. By diagonalizing the Hamiltonian H_0, we obtain the atomic dressed states. For simplicity we assume that $\Omega_{v1}=\Omega_{v2}=\Omega_v$. The dressed states are expressed in terms of the bare atomic states as

$$\begin{aligned}|\pm^{\mu}\rangle &= \tfrac{1}{2}(|1^{\mu}\rangle \pm \sqrt{2}\,|2^{\mu}\rangle + |3^{\mu}\rangle),\\ |d^{\mu}\rangle &= \tfrac{1}{\sqrt{2}}(-|1^{\mu}\rangle + |3^{\mu}\rangle).\end{aligned} \tag{100}$$

Then the Hamiltonian H_0 is rewritten into $\widetilde{H}_0 = \sum_{\mu=1}^{N}\sqrt{2}\hbar\Omega_v(\sigma^{\mu}_{++}-\sigma^{\mu}_{--})$ in the dressed picture. We rewrite the Hamiltonian H_I in terms of dressed states and then transform into the second interaction picture with $\exp(-i\widetilde{H}_0 t/\hbar)$. We assume that the dressed states are well separated from each other. This is guaranteed when Ω_v>>(γ_l,Λ,$|g_l\langle a_l\rangle|$). We tune the cavity fields $\Delta_0=\delta_l=\sqrt{2}\Omega_v$ (l=1,2,3), and make a rotating wave approximation. The Hamiltonian is rewritten as

$$H_I = \sum_{\mu=1}^{N}\hbar\Omega(\sigma^{\mu}_{+4}+\sigma^{\mu}_{4+})+\sum_{\mu=1}^{N}\hbar g(B_1\sigma^{\mu}_{+0}+\sigma^{\mu}_{0+}B_1^{+}), \tag{101}$$

where $g_1=g_2=g_3=g$ has been assumed. Only the sum mode B_1 is coupled to the active medium, and both modes $B_{2,3}$ decouple from the system dynamics. Ω acts as a pumping field on transition $|4^{\mu}\rangle - |+^{\mu}\rangle$. We separate the density operators for the collective modes and obtain the master equation for the sum mode B_1. In the second interaction picture the density operator is related to that in the Schrödinger picture. Since the modes $B_{2,3}$ are not involved in the interaction and are only coupled to a loss reservoir (at zero temperature), they stay in their vacuum states. This leads us to write the total density operator as $\rho = \rho_{B_1}\rho_{B_2}\rho_{B_3}$, where ρ_{B_2} and ρ_{B_3} follow the same equation as Eq. (83). After the transformation, we finally obtain the master equation of the same form as Eq. (84) for the reduced-density operator ρ_{B_1} for mode B_1 and the atoms. The atomic damping term takes the form

$$L_a\rho_{B_1} = \sum_{l=\pm,d}(L_{0l}\rho_{B_1} + L_{l0}\rho_{B_1} + L_{4l}\rho_{B_1}). \tag{102}$$

We assume that κ_1=κ_2=κ_3=κ and so the damping terms for the collective modes have the same form as for the original modes.

In terms of dressed states and the collective modes, the present system is similar to a single-mode system. As a consequence of quantum beats, modes $B_{2,3}$ decouple and only mode B_1 mediates into the interaction. Incoherent transfer of population is carried out through the channel $|0^{\mu}\rangle \xrightarrow{\Lambda} |\pm^{\mu},d^{\mu}\rangle \xrightarrow{\gamma_1} |4^{\mu}\rangle$. Different from the three-level model for lasing without inversion, in the present scheme there are two additional pathways, which involve states $|-^{\mu},d^{\mu}\rangle$. This is advantageous to the population transfer and light amplification. The B_1 mode amplification is realized through the successive coherent driving transition $|4^{\mu}\rangle \xrightarrow{\Omega} |+^{\mu}\rangle$ and lasing transition $|0^{\mu}\rangle \xrightarrow{B_1} |+^{\mu}\rangle$. Through the incoherent and coherent pathways, the laser electron is recycled. We will show that such a mechanism yields three beams of sub-shot light and entanglement among them. From the master equation we can obtain the intensity and the fluctuations of the mode B_1. Using the quantum statistical properties of the collective modes $B_{1,2,3}$, we can obtain the intensities and the quantum statistical properties of the original modes $a_{1,2,3}$.

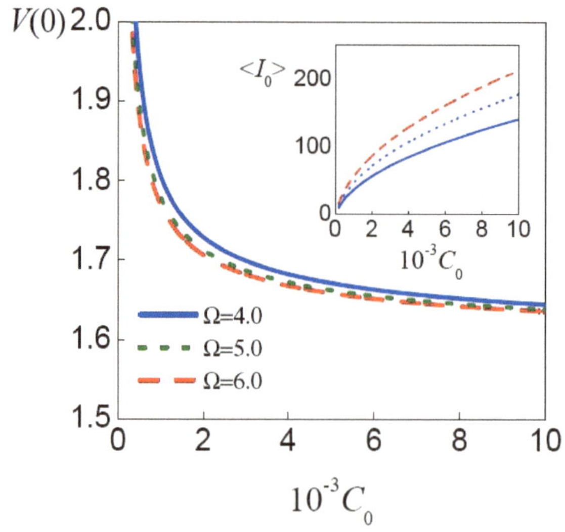

Figure 16: The zero-frequency output spectrum $V(0)$ as a function of the cooperativity parameter C_0 for Ω=4.0 (solid), 5.0 (dotted), and 6.0 (dashed), γ_1=4.0, Λ=0.5, and κ=0.1. Plotted in the inset is the laser intensity $\langle I_0\rangle$ in units of γ_2^2/g^2.

In our numerical calculation, we scale the Rabi frequency Ω and various rates in units of γ_1. The intensities $\langle I_0\rangle = \langle B_1^+ B_1\rangle$ is in units of γ_1^2. Plotted in Fig. 16 is the zero-frequency output spectrum $V(0)$=$V_{213}(0)$= $V_{231}(0)$ as a function of the cooperativity parameter C_0=$2g^2N/(\kappa\gamma_2)$ for Ω=4.0 (solid), 5.0 (dotted), and 6.0 (dashed). The other parameters are chosen as γ_1=4.0, Λ=0.5, and κ=0.1. In the inset we also plot the intensity $\langle I_0\rangle$ for the sum mode B_1. The three generated fields have the characteristic features as follows.

(i) The active device operates well above threshold. The steady state photon number for the sum mode B_1 is of order of hundreds of γ_2^2/g^2, as shown in the inset in Fig. 16. The intensities of respective modes $\langle I_{1,2,3}\rangle$ are of the same order as $\langle I_0\rangle$.

(ii) All of three generated fields have sub-Poissonian statistics. It is seen from Fig. 16 that the output fluctuation spectrum $S(0)= V(0)-2$ for the sum mode B_1 takes its value $S(0)<0$ when the system is operated well above threshold ($C_0 >> 1$). The spectrum S(0) deviates further from zero as C_0 rises. At the same time, the intensity fluctuations depend on the control parameters such as the Rabi frequency Ω. A larger Rabi frequency leads to a smaller value in the spectrum $S(0)$. The best achievable value is $S(0)\approx-\frac{1}{2}$, indicating noise squeezing of 50%. According to Eq. (76), the spectra for all modes are negative, $S_l(0)<0$, which indicates that fluctuations of all of three original cavity fields are suppressed below the shot noise. Quantum-beat and coherence controlled intrinsic feedback combine to cause the squeezing of all three laser modes.

(iii) Three generated fields are entangled. For a wide range of parameters, as shown in Fig. 16, we have $V(0)<2$, which corresponds to the tripartite entanglement. In particular, the existence of the tripartite entanglement is compatible with the laser operation well above threshold. This is remarkably different from the case of the entanglement amplifier [75], where entanglement only appears near above threshold.

Experimentally the three-mode entanglement from the quantum-beat lasers is realizable when one combines the CEL's [41,44] and the lasers without population inversion [54-56], both of which have been realized respectively. Neutral Sn is a possible candidate. The ground-state $|0\rangle=|5s^25p^2\ ^3P_0\rangle$, $|4\rangle=|5s^25p^2_0\ ^1S_0\rangle$, Zeeman sublevels of $|5s^25p5d\ ^1P_1\rangle$ are used as $|1,2,3\rangle$. A magnetic field of 1G splits the $^1P_1$ state into three sublevels, whose spacing is of ~14 GHZ. Quantum-beats between three modes are created by using a microwave field. The microwave frequency is far larger than Rabi frequency, which is much larger than the atomic decay rates. This guarantees that the rotating wave approximations are valid. Excited states decay to state $|4\rangle$ more rapidly than to state $|0\rangle$. This provides the incoherent population transfer from $|0\rangle$ through $|\pm,d\rangle$ to $|4\rangle$. Another possible example is ^{87}Rb D_2 line. We take levels $|0\rangle=|5S_{1/2},F=1\rangle$, and $|4\rangle=|5S_{1/2},F=2\rangle$, which have a spacing of 6.8 GHz. The excited states are $|1\rangle=|5P_{3/2},F=1\rangle$, $|2\rangle=|5P_{3/2},F=2\rangle$, and $|3\rangle=|5P_{3/2},F=3\rangle$. The transitions $|1\rangle-|2\rangle$ and $|2\rangle-|3\rangle$ have frequencies of 157 MHz and 267 MHz, and the excited states have decay rates $(\gamma_1+\gamma_2)/2\pi\leq 6$ MHz. When we use the microwave Rabi frequencies $\Omega_{v1,v2}$~20 MHz, the cross coupling are avoided.

References

1. N. Bloembergen. Nonlinear Optics. New York : Benjamin;1964.
2. Y. R. Shen. The Principles of Nonlinear Optics. New York:Wiley;1984.
3. R. W. Boyd. Nonlinear Optics. Boston :Academic Press;1992.
4. P. Meystre , M. Sargent III. Elements of Quantum Optics 2nd edn. Berlin: Springer;1990.
5. M. O. Scully , M. S. Zubairy. Quantum Optics. Cambridge: Cambridge University Press;1997.
6. R. S. Bondurant, P. Kumar, J. H. Shapiro, M. Maeda. Degenerate four-wave mixing as a possible source of squeezed-state light. Physical Revew A 1984 Jul;30(1):343-353.
7. M. D. Reid, D. F. Walls, and D. J. Dalton.Squeezing of Quantum Fluctuations via Atomic Coherence Effects. Physical Review Letters 1985 Sep 15;55(12):1288-1290.
8. M. D. Levenson, R. M. Shelby, A. Aspect, M. Reid and D. F. Walls. Generation and detection of squeezed states of light by nondegenerate four-wave mixing in an optical fiber. Physical Revew A 1985 Sep;32(3):1550-1562.

9. M. D. Reid and D. F. Walls. Squeezing in nondegenerate four-wave mixing. Physical Revew A 1986 Jun;33(6):4465 -4468.

10. G. S. Agarwal. Generation of Pair Coherent States and Squeezing via the Competition of Four-Wave Mixing and Amplified Spontaneous Emission . Physical Review Letters 1986 Aug 18;57(7):827-830.

11. R. E. Slusher, L. W. Hollberg, B. Yurke, J. C. Mertz, and J. F. Valley. Observation of Squeezed States Generated by Four-Wave Mixing in an Optical Cavity. Physical Review Letters 1985 Nov 25;55(22):788-788.

12. T. Quang. Squeezing via mixing of two modes in a system of driven two-level atoms. Physical Revew A 1990 Jun 1;41(11):6313-6319.

13. M.O. Scully, K. Wodkiewicz, M. S. Zubairy and et al.Two-photon correlated–spontaneous-emission laser: Quantum noise quenching and squeezing. Physical Review Letters 1988 Jun 2;60(18):1832-1835.

14. N. Lu and S. Y. Zhu. Quantum theory of two-photon correlated-spontaneous-emission lasers: Exact atom-field interaction Hamiltonian approach. Physical Revew A 1989 Nov 15; 40(10): 5735-5752.

15. N. Lu and S. Y. Zhu. Quantum theory of a two-mode two-photon correlated spontaneous-emission laser. Physical Revew A 1990 Mar;41(5):2865-2868.

16. J. Bergou, C. Benkert, L. Davidovich and et al. Double two-photon correlated-spontaneous-emission lasers as bright sources of squeezed light .Physical Revew A 1990 Nov 1;42(9):5544-5552.

17. H. J. Kim, A. H. Khosa, H. W. Lee, and M. S. Zubairy. One-atom correlated-emission laser phys. Physical Revew A. 2008 Feb 14;77 (2):023817(9).

18. C. W. Gardiner and P. Zoller. Quantum Noise. Berlin: Springer-Verlag;2000.

19. D. F. Walls , G. J. Milburn. Quantum Optics. Berlin:Springer-Verlag;1994.

20. E.Arimondo. Coherent Population Trapping in Laser Spectroscopy. Elsevier Science 1996;35:257-354

21. S. E. Harris. Laser without inversion: Interference of lifetime-broadened . Physics Today 1997; 50 (7):36-42.

22. J. P. Marangos, Topical review of electromagnetically induced transparency. Journal of Modern Optics 1998;45(5):471-524.

23. M. D. Lukin. Trapping and manipulating photon states in atomic ensembles. Reviews of Modern Physics 2003 Apr;75(2):457 -472.

24. P. W. Milonni. Controlling the speed of light pulses . Journal of Physics B 2002 Mar 13;35:R31-R56

25. G. S. Agarwal and S. P. Tewari. Large enhancements in nonlinear generation by external electromagnetic fields. Physical Review Letters 1993 March 8;70(10):1417-1420.

26. S. Zibrov, M. D. Lukin, and M. O. Scully. Nondegenerate Parametric Self-Oscillation via Multiwave Mixing in Coherent Atomic Media. Phys. Rev. Lett. 1999 November 15;83(20): 4049-4052.

27. L. Deng, M. Kozuma, E. W. Hagley1, and M. G. Payne. Opening Optical Four-Wave Mixing Channels with Giant Enhancement Using Ultraslow Pump Waves. Physical Review Letters. 2002 Apr 8;88(14):143902(3).

28. L. Deng and M. G. Payne. Cherenkov Effect as a Probe of Photonic Nanostructures . Physical

Review Letters 2003 Oct 3;91(14):243902 (4).
29. Y. Wu, J. Saldana, and Y. Zhu. Large enhancement of four-wave mixing by suppression of photon absorption from electromagnetically induced transparency. Physical Revew A 2003 Jan 28;67(1):01381 (5).
30. D. A. Braje, V. Balić, S. Goda, G. Y. Yin, and S. E. Harris. Frequency Mixing Using Electromagnetically Induced Transparency in Cold Atoms. Physical Review Letters 2004 Oct 29;93(18):183601(4).
31. H. Kang, G. Hernandez, and Y. Zhu. Slow-Light Six-Wave Mixing at Low Light Intensities. Physical Review Letters 2004 Au 13;93(7):073601(4).
32. Z. Zuo, J. Sun, X. Liu, Q. Jiang, G. Fu, Ling-An Wu, and P Fu. Generalized n-Photon Resonant 2n-Wave Mixing in an (n+1)-Level System with Phase-Conjugate Geometry. Physical Review Letters 2006 Nov 8;97(16):193904 (4).
33. Y. Zhang, A. W. Brown, and Min Xiao. Opening Four-Wave Mixing and Six-Wave Mixing Channels via Dual Electromagnetically Induced Transparency Windows. Physical Review Letters 2007 Sep 18;99(12): 123603 (4).
34. R. Guzman, J. C. Retamal, E. Solano, and N. Zagury. Field Squeeze Operators in Optical Cavities with Atomic Ensembles. Physical Review Letters 2006 Jan 1;96(1):010502 (4).
35. S. Pielawa, G. Morigi, D. Vitali, and L. Davidovich. Generation of Einstein-Podolsky-Rosen-Entangled Radiation through an Atomic Reservoir. Physical Review Letters 2007 Jun 12;98(24):240401(4).
36. M. Ikram, G. X. Li, and M. S. Zubairy. Entanglement generation in a two-mode quantum beat laser. Physical Revew A 2007 Oct 11;76(4):042317 (6).
37. G. X. Li, H. T. Tan, and M. Macovei. Enhancement of entanglement for two-mode fields generated from four-wave mixing with the help of the auxiliary atomic transition. Physical Revew A 2007 Nov 20;76(5):053827 (9).
38. G. L. Cheng, X. M. Hu, W. X. Zhong, and Q. Li. Two-channel interaction of squeeze-transformed modes with dressed atoms: Entanglement enhancement in four-wave mixing in three-level systems. Physical Revew A 2008 Sep 9;78(3):033811(9).
39. J. Y. Li and X. M. Hu, J. Enhancement of quantum correlations between Rabi sidebands via dressed population transfer . Journal of Physics B 2009 Mar 14;42(5):055501(6).
40. G. L. Cheng, X. M. Hu, and W. X. Zhong. Einstein-Podolsky-Rosen entanglement via nonlinear processes enhanced by electromagnetically induced transparency. Journal of Physics B 2009 Oct 14;42(19):195503(6).
41. S. L. Braunstein and P. van Loock. Quantum information with continuous variables Reviews of Modern Physics 2005 Apr;77(2):513-577.
42. A. Einstein, B, Podolsky, and N. Rosen. Can Quantum-Mechanical Description of Physical Reality Be Considered Complete? Physical Revew 1935 May 15;47 (10):777 -780.
43. M. O. Scully. Correlated Spontaneous-Emission Lasers: Quenching of Quantum Fluctuations in the Relative Phase Angle. Physical Review Letters 1985 Dec 16;55(25):2802-2805.
44. M. O. Scully and M. S. Zubairy. Theory of the quantum-beat laser. Physical Revew A 1987 Jan 15;35(2):752-759.
45. J. Bergou, M. Orszag, and M. O. Scully. Correlated-emission laser: Phase noise quenching via coherent pumping and the effect of atomic motion . Physical Revew A 1988 Jul 15;38 (2): 768-772.

46. M. P. Winters, J. L. Hall, and P. E. Toschek. Correlated spontaneous emission in a Zeeman laser. Physical Review Letters 1990 Dec 17;65(25):3116-3119.

47. N. Lu. Quantum theory of correlated-emission lasers: Vacuum state for the mode of the relative phase and the relative amplitude. Physical Revew A 1992 Jun 1;45(11):8154-8164.

48. U. W. Rathe and M. O. Scully. Phase coherence and decoherence in the correlated-spontaneous-emission laser. Physical Revew A 1995 Oct;52 (4):3193-3200.

49 Ingo Steiner, Peter E. Toschek Quenching Quantum Phase Noise: Correlated Spontaneous Emission versus Phase Locking . Physical Review Letters 1995 Jun 5;74(23):4639 – 4642.

50. O. Kocharovskaya. Amplification and lasing without inversion. Physics Reports 1992 Oct;219(3-6):175-190.

51. M. O. Scully. From lasers and masers to phaseonium and phasers. Physics Reports 1992 Oct ;219(3-6):191-192.

52. P. Mandel. Lasing without inversion: a useful concept? Contemporary Physics 1993;34(5), 235-246.

53. Mompart J, Corbalan R Lasing without inversion Journal of Optics B-quantum and Semiclassical Optics 2000 Jun;2 (3):R7-R24.

54. A. S. Zibrov, M. D. Lukin, D. E. Nikonov, L. Hollberg, M. O. Scully, V. L. Velichansky,H. G. Robinson. Experimental Demonstration of Laser Oscillation without Population Inversion via Quantum Interference in Rb . Physical Review Letters 1995 Aug 21;75(8):1499 – 1502.

55. Haibin Wu, Min Xiao, J. Gea-Banacloche. Evidence of lasing without inversion in a hot rubidium vapor under electromagnetically-induced-transparency conditions. Physical Revew 2008 Oct 23 A ;78(4):041802(4).

56. Haibin Wu, Min Xiao, J. Gea-Banacloche. Evidence of lasing without inversion in a hot rubidium vapor under electromagnetically-induced-transparency conditions. Physical Revew A 2008 Oct 23;78(4):041802(4).

57. Klaus M. Gheri , Daniel F. Walls. Lossless transformation of coherent light into amplitude-squeezed light . Physical Revew A 1992 Dec 1;46(11):R6793 - R6796.

58. H. Ritsch, M. A. M. Marte. Quantum noise in Raman lasers: Effects of pump bandwidth and super- and sub-Poissonian pumping . Physical Revew A 1993 Mar;47(6):2354 - 2365

59. Yifu Zhu, Min Xiao. Amplitude squeezing and a transition from lasing with inversion to lasing without inversion in a four-level laser. Physical Revew A 1993 Nov;48(5):3895 – 3899.

60. K. J. Schernthanner, H. Ritsch. Quantum-noise reduction in Raman lasers: Effects of collisions, population trapping, and Doppler shifts. Physical Revew A 1994 May;49(5):4126-4133.

61. Carlos Saavedra, Juan C. Retamal, Christoph H. Keitel. Strong intracavity and output laser noise reduction via initial atomic coherence. Physical Revew A 1997 May;55(5): 3802-3812.

62. Alexander Khazanov, Genady Koganov, Reuben Shuker. Laser-noise suppression in the dressed-atom approach. I. Fluctuations in a regularly pumped laser. Physical Revew A 1993 Jun; 48(2):1661-1670.

63. H. Ritsch, P. Zoller. Dynamic quantum-noise reduction in multilevel-laser systems. Physical Revew A 1992 Aug;45(3):1881-1892.

64. T. C. Ralph, C. M. Savage. Squeezed light from a coherently pumped four-level laser. Physical Revew A 1991 Dec 1 ; 44(11):7809-7814.

65. D. L. Hart, T. A. B. Kennedy. Quantum noise properties of the laser: Depleted pump regime. Physical Revew A 1991 Oct 1;44(7):4572 – 4577.

66. X. M. Hu and J. S. Peng. Dynamic quantum noise reduction in a Λ quantum-beat laser. Optics Communications 1998 Aug 15;154(1-3):152-159.
67. X. M. Hu and J. S. Peng. Sub-Poissonian quantum-beat laser without inversion. Journal of Physics B 1998 Dec 28; 31(24):5393-5402.
68. X. M. Hu and J. S. Peng. Squeezed two-mode lasers without and with inversion. The European Physical Journal D 1999 Feb;5(2):291-299.
69. H. Xiong, M. O. Scully, M. S. Zubairy. Correlated Spontaneous Emission Laser as an Entanglement Amplifier. Physical Review Letters 2004 Jul 21;94(2):023601(4).
70. H. T. Tan, S. Y. Zhu, M. S. Zubairy. Continuous-variable entanglement in a correlated spontaneous emission laser. Physical Revew A 2005 Aug 8;72(2):022305(8).
71. Ling Zhou, Han Xiong, M. Suhail Zubairy Single atom as a macroscopic entanglement source. Physical Revew A 2006 Oct 12;74(2):022321 (5).
72. M. Kiffner,M. S. Zubairy,J. Evers, C. H. Keitel. Two-mode single-atom laser as a source of entangled light. Physical Revew A 2007 Jul 2;75(3):033816(8).
73. Shahid Qamar, Han Xiong, M. Suhail Zubairy. Influence of pump-phase fluctuations on entanglement generation using a correlated spontaneous-emission laser. Physical Revew A 2007 Oct 3; 75(6):062305(10).
74. Shahid Qamar, Fazal Ghafoor, Mark Hillery, M. Suhail Zubairy. Quantum beat laser as a source of entangled radiation. Physical Revew A 2008 Jun 16; 77(6):062308(7).
75. W. X. Shi, X. M. Hu, and F. Wang. Three-mode entanglement and amplification in correlated spontaneous emission lasers. Journal of Physics B 2009 Aug 28;42(16):165506.
76. Xiang-ming Hu and Jin-hua Zou. Quantum-beat lasers as bright sources of entangled sub-Poissonian light. Physical Revew A 2008 Sep 15;78(4):045801(4).
77. X. Y. Lü, J. B. Liu, L. G. Si and et al. Continuous-variable entanglement in a two-mode four-level single-atom laser. Journal of Physics B 2008 Feb 14;41(3):035501.
78. X. M. Hu and J. H. Zou. Quantum-beat lasers as bright sources of entangled sub-Poissonian light. Physical Revew A 2008 Oct 8; 78(4): 045801(4).
79. Xiaoxia Li and Xiangming Hu. Tripartite entanglement in quantum-beat lasers. Physical Revew A 2009 Jun 1; 80(2): 023815(8).
80. Lu-Ming Duan, G. Giedke, J. I. Cirac.Inseparability Criterion for Continuous Variable Systems. Physical Review Letters 2000 Oct 5; 84(12):2722 – 2725.
81. J. S. Peng and G. X. Li. Introduction to Modern Quantum Optics. Singapore:World Scientific; 1998.
82. H. Haken. Laser Theory, in Encyclopedia of Physics. Berlin: Springer-Verlag;1970.
83. W. H. Louisell. Quantun Statistical Properties of Radiation. New York: Wiley ; 1973.
84. M. Sargent III, M. O. Scully, and W. E. Lamb, Jr.. Laser Physics, MA :Addison-Wesley; 1974.
85. C. Cohen-Tannoudji, J. Dupont-Roc and G. Grynberg, Atom-Photon. Interactions-Basic Processes and Applications .The Physical Society of Japan 1993;48(9):745-746.
86. E. Solano, R. L. de Matos Filho, N C. Cohen-Tannoudji, J. Dupont-Roc, G. Grynberg, Atom-Photon. Interactions-Basic Processes and Applications 1992. Zagury Mesoscopic Superpositions of Vibronic Collective States of N Trapped Ions. Physical Review Letters 2001 Aug 6; 87(3):060402(4).
87. E. Solano, G. S. Agarwal, H. Walther Strong-Driving-Assisted Multipartite Entanglement in Cavity QED. Physical Review Letters 2003 Jun 17;90(11): 027903(4).

88. P. Lougovski, P. E. Solano, H. Walther. Generation and purification of maximally entangled atomic states in optical cavities. Physical Revew A 2005 Jan 28; 71(1):013811(4).
89. X. L. Feng, Z. S. Wang, C. F. Wu, L. C. Kwek, C. H. Lai, and C. H. Oh. Scheme for unconventional geometric quantum computation in cavity QED. Physical Revew A 2007 May 9;75(5):052312(7).
90. A. S. Zibrov , A. B. Matsko, M. O. Scully. Four-Wave Mixing of Optical and Microwave Fields. Physical Review Letters 2002 Sep 2;89(12):103601(4).
91. Y. Zhao, C. K. Wu, B. S. Ham, M. K. Kim, E. Awad.Microwave Induced Transparency in Ruby. Physical Review Letters 1997 Mar 5;79(4):641-644.
92. D. McGloin ,M. H. Dunn. Simple theory of microwave induced transparency in atomic and molecular systems. Journal of Modern Optics 1999 Sep;47(11): 1887-1897.
93. S. F. Yelin, V. A. Sautenkov, M. M. Kash, G. R. Welch, M. D. Lukin. Nonlinear optics via double dark resonances. Physical Revew A 2003 Dec 2;68(6):063801(7).
94. M. Jain, H. Xia, G. Y. Yin, A. J. Merriam, S. E. Harris. Efficient Nonlinear Frequency Conversion with Maximal Atomic Coherence. Physical Review Letters 1996 Jun 14; 77(21):4326-4329.
95. J. Q. Liang, M. Katsuragawa, F. L. Kien, K. Hakuta. Sideband Generation Using Strongly Driven Raman Coherence in Solid Hydrogen. Physical Review Letters 2000 Mar 7;85(12): 2474-2477.
96. S. E. Harris, J. E. Field,A. Imamoglu. Nonlinear optical processes using electromagnetically induced transparency. Physical Review Letters 1990 Dec 27;64(10):1107-1110.
97. K. M. Gheri, C. Saavedra, D. F. Walls. Intracavity second-harmonic generation using an electromagnetically induced transparency. Physical Revew A 1993 May 4; 48(4):3344-3361.
98. K. Hakuta, L. Marmet, B. P. Stoicheff. Electric-field-induced second-harmonic generation with reduced absorption in atomic hydrogen. Physical Review Letters 1991 Aug 6;66(5): 596-599 .
99. G. Z. Zhang, K. Hakuta, B. P. Stoicheff. Nonlinear optical generation using electromagnetically induced transparency in atomic hydrogen. Physical Review Letters 1993 Aug 17 ;71(19):3099-3102.
100. G. Z. Zhang, M. Katsuragawa, K. Hakuta, R. I. Thompson, B. P. Stoicheff. Sum-frequency generation using strong-field coupling and induced transparency in atomic hydrogen. Physical Revew A 1995 Mar 7;52(2):1584-1593.
101. G. Z. Zhang, D. W. Tokaryk, B. P. Stoicheff, K. Hakuta. Nonlinear generation of extreme-ultraviolet radiation in atomic hydrogen using electromagnetically induced transparency. Physical Revew A 1997 Dec 2;56(1):813-819.
102. M. Oberst , J. Klein, and T. Halfmann. Enhanced four-wave mixing in mercury isotopes, prepared by stark-chirped rapid adiabatic. Optics Communications 2006 Aug 15;264(2):463-470.
103. L. Davidovich. Sub-Poissonian processes in quantum optics. Reviews of Modern Physics 1996 Jan;68(1):127-173.
104. P. van Loock S. L. Braunstein. Multipartite Entanglement for Continuous Variables: A Quantum Teleportation Network. Physical Review Letters 1999 Jun 17;84(15):3482-3485.
105. M. Murao, D. Jonathan, M. B. Plenio, V. Vedral. Quantum telecloning and multiparticle entanglement. Physical Revew A 1999 Jun 24;59(1):156-159.
P. van Loock ,S. L. Braunstein. Telecloning of Continuous Quantum Variables. Physical Review

Letters 2001 Nov 26; 87(24):247901(4).

106. J. Zhang, C. D. Xie, K. C. Peng. Controlled dense coding for continuous variables using three-particle entangled states. Physical Revew A 2002 Sep 26;66(3):032318(6).

J. T. Jing, J. Zhang, Y. Yan, F. G. Zhao, C. D. Xie, K. C. Peng. Experimental Demonstration of Tripartite Entanglement and Controlled Dense Coding for Continuous Variables. Physical Review Letters 2003 Apr 23;90(16);167903(4).

107. D. M. Greenberger, M. A. Horne, A. Shimony, and A. Zeilinger. American Journal of Physics 1990 58, 1131; D. M. Greenberger, M. A. Horne, and A. Zeilinger, Physics Today 46 (8), 22 (1993)

108. J. Guo, H. X. Zou, Z. H. Zhai, J. X. Zhang, J. R. Gao. Generation of continuous-variable tripartite entanglement using cascaded nonlinearities. Physical Revew A 2005 Mar 15; 71(3):034305 (4).

109. A. S. Villar, M. Martinelli, C. Fabre, P. Nussenzveig. Direct Production of Tripartite Pump-Signal-Idler Entanglement in the Above-Threshold Optical Parametric Oscillator. Physical Review Letters 2006 Oct 6;97(14):140504(4).

110. T. Aoki, N. Takei, H. Yonezawa, K. Wakui, T. Hiraoka, A. Furusawa, P. van Loock. Experimental Creation of a Fully Inseparable Tripartite Continuous-Variable State. Physical Review Letters 2003 Aug 21;91(8):80404(4) .

111. M. K. Olsen, A. S. Bradley, and M. D. Reid. Continuous variable tripartite entanglement and Einstein–Podolsky–Rosen correlations from triple nonlinearities. Journal of Physics B 2006 Mar 2;39:2515-2533.

112. X. L. Su, A. H. Tan, X. J. Jia, J. Zhang, C. D. Xie, K. C. Peng. Experimental Preparation of Quadripartite Cluster and Greenberger-Horne-Zeilinger Entangled States for Continuous Var. Physical Review Letters 2007 Feb 14;98(7):070502(4) .

113. P. van Loock and A. Furusawa. Detecting genuine multipartite continuous-variable entanglement. Physical Revew A 2003 May 29;67(5):052315(13).

114. M. K. Olsen A. S. Bradley. Asymmetric polychromatic tripartite entanglement from interlinked χ (2) parametric interactions. Physical Revew A 2006 Dec 11;74(6):063809(8).

115. P. D. Drummond, C. W. Gardiner. Generalised P-representations in quantum optics. Journal of Physics A 1980 Jan 2;13:2353-2368.

P. D. Drummond ,D. F. Walls. Quantum theory of optical bistability. II. Atomic fluorescence in a high-Q cavity. Physical Revew A 1981 Jul 7;23(5):2563-2579.

116. K. Bergmann, H. Theuer, B. W. Shore. Coherent population transfer among quantum states of atoms and molecules. Reviews of Modern Physics 1998 Jul;70(3):1003-1025.

117. P. Král, I. Thanopulos, and M. Shapiro. Colloquium: Coherently controlled adiabatic passage. Reviews of Modern Physics 2007 Jan 2;9(1):53(25).

Chapter 5. Tunable Photonic Bandgaps Induced by Standing-Wave Fields and Related Topics

Jin-Hui Wu[1], M. Artoni[2,3], and G. C. La Rocca[4], Jin-Yue Gao[1]
1) Jilin University
2) Brescia University
3) European Laboratory for Nonlinear Spectroscopy
4) Scuola Normale Superiore and CNISM

Abstract: In this chapter, we give a brief review of our recent research works on photonic bandgaps induced by standing-wave (SW) coupling fields in the regime of electromagnetically induced transparency (EIT). EIT refers to the absorption suppression or elimination of a weak probe going through an atomic ensemble at the presence of a strong coupling, which is a result of laser induced destructive quantum interference. On the other hand, a light signal cannot freely propagate in the media with periodic refractive indices (the so-called photonic crystals) if its carrier frequencies fall inside the photonic bandgaps. Utilizing a SW coupling to attain the periodically modulated refractive index with little absorption, one can establish an induced photonic bandgap with its width and position dynamically tunable. The potential media may be either cold atomic ensembles or solid materials exhibiting defect states, such as Pr3+: Y2SiO5 and diamond containing N-V color centers. We first consider the steady optical responses of the atomic and solid media with dynamically induced bandgaps to a time- independent probe by focusing on the dispersion curves of Bloch wave vectors and the spectra of reflection and transmission. Then, we examine the propagation dynamics of a probe pulse through a cold atomic sample in two different situations where a dynamically induced bandgap exists or not. We find that the SW coupling field can be easily modulated to mold the traveling-light flow and to control the stationary-light generation with quite high flexibility, and thus may have applications in the fields of classical and quantum information processing of optical signals. In performing theoretical simulations, we also demonstrated several different but almost equivalent mathematical methods, i.e. the two-mode approximation method, the transfer-matrix method, and the Maxwell-Liouville equation method, to deal with the problems of SW-EIT. Comparison of these methods is quite helpful to understand the underlying physics of the formation of dynamically induced photonic bandgaps.

Introduction

Quantum coherence induced by laser fields in atomic gases and impurity doped solids lies at the heart of quantum optics, especially for their applications to light wave propagation control, enhancement of resonant optical nonlinearity and state transfer between light and atoms, just to mention a few. The main impetus for recent intense activities on laser induced quantum coherence comes from its potential applications in quantum information storage and processing with photonic wave packets. One relevant phenomenon (maybe the most important one) is electromagnetically induced transparency (EIT) [1-4] referring to the absorption suppression or elimination of a weak probe field in certain media due to the destructive quantum coherence induced by a moderate coupling field when the necessary condition of two-photon resonance is satisfied. With a *static* cw coupling, one can easily slow down a probe pulse to several meters per second with little loss utilizing the extremely steep dispersion of refractive index at the center of an EIT window [5, 6]. By a *dynamic* modulation of the cw coupling in time one can instead transfer quantum states of the probe pulse to low-frequency (spin) coherences of the dressed media (atomic gases or impurity doped solids) and retrieve it after a short time [7-9]. Storage times in this case are usually of the order of milliseconds and mainly limited by the dephasing rates of low-frequency (spin) coherences. Quite recently, this light manipulation technique has been extended to demonstrate the storage and retrieval of single photons [10], the photonic entanglement [11], and the squeezed vacuum [12], which is surely an essential step toward its real applications in quantum computation and communications. Slightly disturbing the typical Lambda-type EIT system through an auxiliary field on an auxiliary transition, one may further attain the greatly enhanced Kerr-type optical nonlinearity with little loss [13-15], which has been exploited to devise an efficient quantum phase gate [16, 17] and a novel all-optical switching [18, 19], etc.

An important step toward light storage has been taken by employing a *standing-wave* (*SW*) coupling field instead of the *traveling-wave* (*TW*) coupling field in typical Lambda-type EIT systems. Using a perfect *SW* coupling with equal strengths of forward (FW) and backward (BW) components, stationary light pulses (serving as stored signals) were successfully generated in experiment from low-frequency (spin) coherences of warm [20] and cold [21] atomic samples. Such stationary light signals may be used, e.g., to devise interesting new schemes to enhance the Kerr-like nonlinear optical interaction [22] or to achieve quantum transport of single photons [23]. Detailed theoretical treatments on the stationary light generation with a perfect *SW* coupling are given in terms of the coupled *Maxwell-Liouville* equations as in Refs. [24, 25], where the secular or adiabatic approximations have been made. However, the different optical responses of warm atoms (with Doppler broadening) and of cold atoms (without Doppler broadening) driven by a perfect *SW* coupling [26] are not very clear yet as far as the propagation dynamics of a probe pulse is concerned, which will be emphasized here.

Another important issue relevant to the SW-EIT scheme is that a photonic bandgap is expected to open up in the EIT window because the absorptive and dispersive properties on the probe resonance are periodically modulated in a controllable way. To well eliminate the probe absorption at the *SW* nodes and then have a rather good photonic bandgap, however, one should use an imperfect *SW* coupling with unequal FW and BW components on cold atomic clouds or impurity doped solids such as Pr^{3+} doped Y_2SiO_5 crystal and diamond containing nitrogen- vacancy (N-V) color centers [27-30]. The probe absorption, as a obstacle to the photonic bandgap generation, may also be avoided by exploiting the far-off-resonance quantum coherence induced by a perfect *SW* coupling [31, 32]. It is certainly of great interest that the coherently induced photonic bandgaps can be generated and destroyed on demand and can be expediently modified in terms of their widths and positions. The corresponding reflectivity can never reach 100% due to the residual absorption in the EIT window and is usually over 90% for atomic or solid samples of realistic density and length. Splitting a periodically modulated EIT window by introducing a second *TW* coupling, one may further establish two adjacent photonic bandgaps in cold atoms with four energy levels. Other optical methods, e.g. through the AC Stark effect and the coherently enhanced refractive index, have also been proposed to attain controllable photonic bandgaps in semiconductor quantum wells [33, 34]. Photonic bandgaps induced by external fields make it clearly unnecessary to design photonic crystals [35] with predetermined bandgap structures, and thus may be very useful for controlling the light flow and interactions.

In this chapter, we give a brief review of our recent investigations on the steady and dynamical optical responses of cold ^{87}Rb atoms and N-V color centers in diamond driven by a *SW* coupling in the Lambda configuration. In section II, we focus on the steady responses of these media to a static (cw) probe in terms of the Bloch wavevector and the reflection and transmission spectra. Starting from the dressed susceptibility of cold ^{87}Rb atoms, we first derive the Bloch eigenstates of electric and magnetic probe fields within a two-mode approximation from which analytical expressions for the bandgap structure and relevant spectra can be obtained. With appropriate parameters, we can observe a well developed bandgap (about 0.5 MHz of width) and a high reflectivity (about 95% of magnitude) of the incident cw probe, which depend critically on Rabi frequencies and other parameters of the *SW* coupling. Then we extend this work to the N-V color centers in diamond by the standard transfer-matrix method where a realistic Gaussian profile of inhomogeneous broadening on the probe resonance has been taken into account. Our numerical calculations show that the induced photonic bandgap may has a width of 20 MHz and the reflectivity therein may exceed 99% in an mm long sample. The nearly perfect reflection in an impurities doped diamond crystal is a direct result of the much better suppressed absorption in the EIT window.

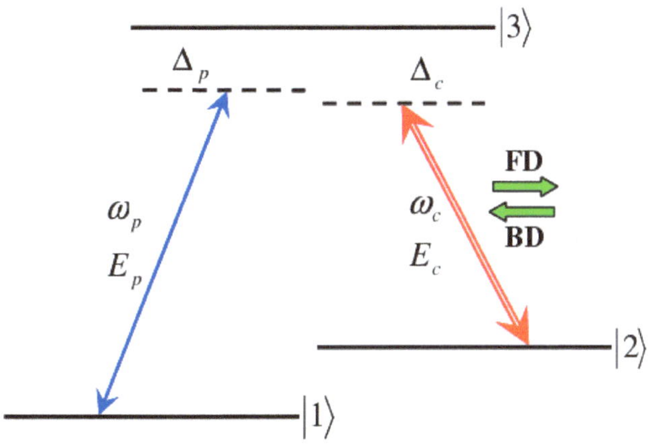

Fig. (1) (color online) Schematic diagram of a three-level atom interacting with a weak probe (*blue*) field and a strong coupling (*red*) field. The coupling is retro-reflected upon impinging on a mirror to form a standing-wave pattern within the atomic sample. A misalignment θ between the forward and backward beams of the standing-wave coupling modifies the lattice periodicity as $\lambda_c/2 \rightarrow \lambda_c/2\cos(\theta/2)$.

In section III, instead we pay attention to the dynamical responses of cold ^{87}Rb atoms in two different situations where the *SW* coupling is either imperfect or perfect. In the first situation, as a photonic bandgap is induced by a imperfect *SW* coupling, a probe pulse can be either perfectly reflected with a shorter time delay or partially transmitted with a longer time delay depending on whether its carrier frequencies fall inside bandgap or not. Here the spin and optical coherences are spatially expanded in terms of the coupling wave vector, which then requires that the Maxwell-Liouville equations be truncated at a proper point of the (spatial) Fourier expansion. We also investigate the significant effect of probe penetration and find that the field distribution may be quite complicated due to the interference between the FW and BW probe components. In the second situation, we further examine the nonlinear generation of a stationary light with the same Maxwell-Liouville equations. It is found that the stationary light generated form a wave packet of atomic spin coherence diffuses and decays quickly as the Maxwell-Liouville equations are properly truncated, e.g. at |n|=40. If the Maxwell-Liouville equations are purposely truncated at |n|=0, however, the stationary light experiences little loss and diffusion due to the unphysically omission of higher-order spin and optical coherences.

Steady optical responses of periodically modulated media

Photonic bandgaps in ultracold atoms driven by a SW light

As shown in Fig. 1, we first consider a three-level Λ- type atomic system interacting with a *probe* of frequency ω_p and wave vector k_p and a coupling of frequency ω_c and wave vector k_c. In the interaction picture, with the rotating-wave and electric-dipole approximations, the interaction Hamiltonian can be written as

$$H_{\text{int}} = -\hbar\Omega_p e^{-i\Delta_p t}|3\rangle\langle 1| - \hbar\Omega_c e^{-i\Delta_c t}|3\rangle\langle 2| + h.c. \tag{1}$$

with $\Delta_p = \omega_p - \omega_{31}$ and $\Delta_c = \omega_c - \omega_{32}$ being detunings of the probe and coupling fields while $\Omega_p = E_p d_{31}/2\hbar$ and $\Omega_c = E_c d_{32}/2\hbar$ denoting the corresponding Rabi frequencies. If we describe the population decay rates by Γ_{31}, Γ_{32}, and Γ_{21} while the coherence dephasing rates by $\gamma_{12} = \Gamma_{21}/2$, $\gamma_{13} = (\Gamma_{31} + \Gamma_{32})/2$, and $\gamma_{23} = (\Gamma_{31} + \Gamma_{32} + \Gamma_{21})/2$, the dynamical evolution equations for the density operator may be expanded as

$$\begin{aligned}
\partial_t \rho_{11} &= \Gamma_{31}\rho_{33} + \Gamma_{21}\rho_{22} + i\Omega_p^*\rho_{31} - i\Omega_p\rho_{13} \\
\partial_t \rho_{22} &= \Gamma_{32}\rho_{33} - \Gamma_{21}\rho_{22} + i\Omega_c^*\rho_{32} - i\Omega_c\rho_{23} \\
\partial_t \rho_{21} &= -[\gamma_{12} - i(\Delta_p - \Delta_c)]\rho_{21} - i\Omega_p\rho_{23} + i\Omega_c^*\rho_{31} \\
\partial_t \rho_{31} &= -[\gamma_{13} - i\Delta_p]\rho_{31} + i\Omega_c\rho_{21} - i\Omega_p(\rho_{33} - \rho_{11}) \\
\partial_t \rho_{32} &= -[\gamma_{23} - i\Delta_c]\rho_{32} + i\Omega_p\rho_{12} - i\Omega_c(\rho_{33} - \rho_{22})
\end{aligned} \tag{2}$$

Solutions of Eqs. (2) then yield the macroscopic polarization $P = Nd_{13}\rho_{31}$ with N being the atomic density and the macroscopic susceptibility $\chi_p = P/(\varepsilon_0 E_p)$ in the limit of a weak probe.

For a resonant coupling ($\Delta_c = 0$), the steady state probe susceptibility can be obtained as

$$\chi_p = \frac{N|d_{13}|^2}{2\hbar\varepsilon_0} \frac{\Delta_p + i\gamma_{12}}{(\gamma_{12} - i\Delta_p)(\gamma_{13} - i\Delta_p) + |\Omega_c|^2} \tag{3}$$

which may generate a steep dispersion within a narrow EIT window centered at $\Delta_p = 0$ if the coupling is in the *TW* configuration.

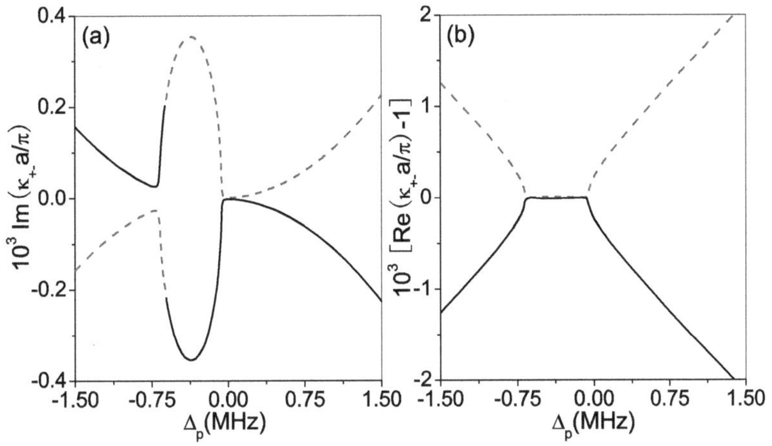

Fig. (2) (color online) Photonic band-gap structures in ultracold ^{87}Rb atoms obtained from Eq. (11). The black-solid and red-dashed curves correspond to κ_+ and κ_-, respectively. Relevant parameters are $\Gamma_{31} = \Gamma_{32} = 6.0MHz$, $\Gamma_{21} = 2.0kHz$, $N = 10^{13} cm^{-3}$, $\theta = 45mrad$, $\Omega_{c+} = 30MHz$, $\Omega_{c-} = 25MHz$, $\lambda_{31} = 780.792nm$, and $\lambda_{32} = 780.778nm$.

When the coupling is in the *SW* configuration, however, Ω_c in Eq. (3) should be replaced by $\Omega_c = \Omega_{c+}e^{ik_cz} + \Omega_{c-}e^{-ik_cz}$ with real Ω_{c+} and Ω_{c-} representing Rabi frequencies of the forward (*FD*) and backward (*BD*) components. Thus the probe susceptibility becomes a periodically modulated function of the form

$$\chi_p = \frac{N|d_{13}|^2}{2\hbar\varepsilon_0} \frac{A}{1 + B\cos(2k_cz)} \tag{4}$$

With

$$A = \frac{\Delta_p + i\gamma_{12}}{\Omega_{c+}^2 + \Omega_{c-}^2 - \Delta_p^2 - i(\gamma_{12} + \gamma_{13})\Delta_p + \gamma_{12}\gamma_{13}}$$
$$B = \frac{2\Omega_{c+}\Omega_{c-}}{\Omega_{c+}^2 + \Omega_{c-}^2 - \Delta_p^2 - i(\gamma_{12} + \gamma_{13})\Delta_p + \gamma_{12}\gamma_{13}} \tag{5}$$

which may be further expanded as a cosine Fourier series [36]

$$\chi_p(z) = \chi_0 + 2\sum_{n=1}^{\infty} \chi_n \cos(2nk_c z) \tag{6}$$

With

$$\chi_n = \frac{k_c}{\pi}\frac{NA|d_{13}|^2}{2\hbar\varepsilon_0}\int_0^{\pi/k_c} \frac{\cos(2nk_c z)dz}{1 + B\cos(2k_c z)}$$
$$= \frac{NA|d_{13}|^2}{2\hbar\varepsilon_0}\frac{1}{\sqrt{1-B^2}}\left(\frac{\sqrt{1-B^2}-1}{B}\right)^n \tag{7}$$

The eigenfunction of an electromagnetic field in a medium described by Eq. (6) can be cast [35] into the form

$$E(z) = \varepsilon(z)e^{i\kappa z} = \left(\sum_{n=-\infty}^{\infty} \varepsilon_n e^{i2nk_c z}\right)e^{i\kappa z} \tag{8}$$

where $\varepsilon(z)$ represents the complex amplitude of the probe electric field polarized along x and propagating along z with the Bloch wave vector κ. For Bloch wave vectors κ around k_c, the electric field in Eq. (8) is essentially determined by the two amplitudes ε_0 and ε_{-1} characterizing the main Bragg components.

The Maxwell wave equation for a monochromatic probe of frequency $\omega_p = ck_p$ in a medium with the susceptibility $\chi_p(z)$ is

$$\frac{\partial^2 E(z)}{\partial z^2} + k_p^2[1 + \chi_p(z)]E(z) = 0 \tag{9}$$

which, when Eq. (6) is inserted into along with ε_0 and ε_{-1} (*two-mode approximation*), reduces to a system of two coupled equations involving only the susceptibility components χ_0 and χ_1. In the matrix form this reads as

$$\begin{pmatrix} k_p^2(1+\chi_0)-\kappa^2 & k_p^2\chi_1 \\ k_p^2\chi_1 & k_p^2(1+\chi_0)-(\kappa-2k_c)^2 \end{pmatrix}\begin{pmatrix} \varepsilon_0 \\ \varepsilon_{-1} \end{pmatrix}=0 \tag{10}$$

It is convenient to set $\kappa = k_c + q$ with q exploring a small region around the Brillouin zone boundary k_c. To the lowest order in q one has

$$q_{\pm} \cong \pm\frac{1}{2k_c}\sqrt{[k_p^2(1+\chi_0)-k_c^2]^2 - k_p^4\chi_1^2} \equiv \pm q \tag{11}$$

which is got from the determinant of Eq. (10) under the realistic assumption that $|\gamma_1| << |1+\chi_0|$ and when half of the probe wavelength inside the medium matches the periodicity. This yields two degenerate Bloch eigenstates of the electromagnetic field with *electric field* eigenfunctions $E^{\pm}(z)$ in the form

$$\begin{aligned} E^+(z) &= \varepsilon_0^+ e^{i\kappa_+ z} + \varepsilon_{-1}^+ e^{-i\kappa_- z} \\ E^-(z) &= \varepsilon_0^- e^{i\kappa_- z} + \varepsilon_{-1}^- e^{-i\kappa_+ z} \end{aligned} \tag{12}$$

corresponding to the two Bloch wave vectors $\kappa_{\pm} = k_c \pm q$ with eigenvectors $\varepsilon^{\pm}_{\{0,-1\}}$.

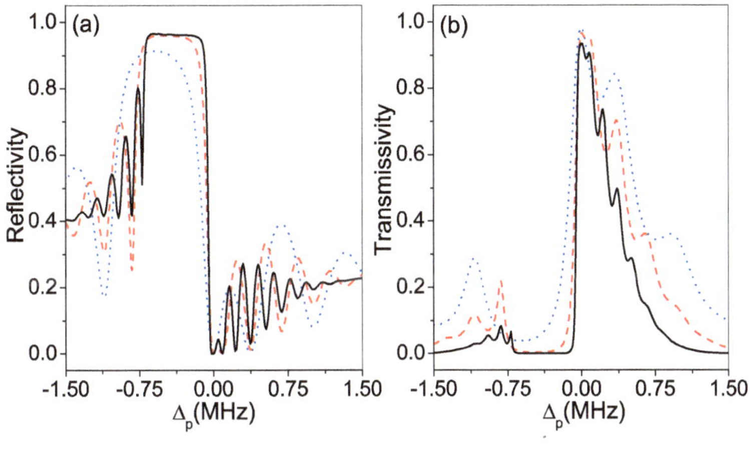

Fig. (3) (color online) Probe reflectivity and transmissivity in an ultracold ^{87}Rb sample of width L=2.0 mm (black-solid), L=1.0 mm (red-dashed) and L=0.5 mm (blue-dotted). Other parameters are the same as in Fig. 2.

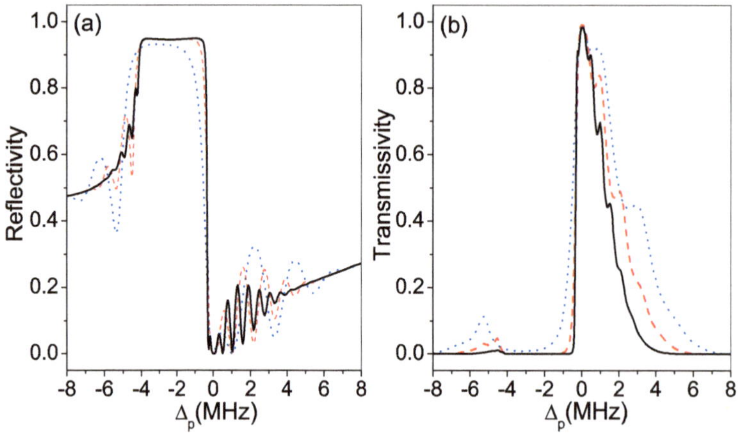

Fig. (4) (color online) Probe reflectivity and transmissivity in an ultracold ^{87}Rb sample of width L=2.0 mm (black-solid), L=1.0 mm (red-dashed) and L=0.5 mm (blue-dotted). Other parameters are the same as in Fig. 2 except $\Omega_{c+} = 60MHz$ and $\Omega_{c-} = 50MHz$.

The electric field in Eq. (12) is seen to combine both FW and BW plane-waves weighted by the corresponding eigenvectors. In fact, the off-diagonal term in Eq. (10) proportional to χ_1 mixes into the Bloch eigenstates, respectively, the plane-wave $e^{i\kappa_+ z}$ with $e^{-i\kappa_- z}$, and the plane-wave $e^{i\kappa_- z}$ with $e^{-i\kappa_+ z}$. Furthermore, with the help of $\partial_z E_p = -(1/c)\partial_t B_p$, one has for the corresponding magnetic field eigenfunctions

$$\begin{aligned} B^+(z) &= (\varepsilon_0^+ \kappa_+ e^{i\kappa_+ z} - \varepsilon_{-1}^+ \kappa_- e^{-i\kappa_- z})/k_p \\ B^-(z) &= (\varepsilon_0^- \kappa_- e^{i\kappa_- z} - \varepsilon_{-1}^- \kappa_+ e^{-i\kappa_+ z})/k_p \end{aligned}, \tag{13}$$

while the total electromagnetic field may be written as

$$\begin{aligned} E_p(z) &= \alpha E^+(z) + \beta E^-(z) \\ B_p(z) &= \alpha B^+(z) + \beta E^-(z) \end{aligned} \tag{14}$$

with α and β to be determined.

For a sample of finite length L, typically containing a very large number of periods a, the reflection and transmission complex amplitudes r and t may be connected to the electric and magnetic fields at the inner (z=0) and outer (z=L) boundaries of the sample through the relations

$$\begin{aligned} E_p(0) &= (1+r)E_{in}(0) \\ B_p(0) &= (1-r)B_{in}(0) \\ E_p(L) &= tE_{in}(0) \\ B_p(L) &= tB_{in}(0) \end{aligned} \tag{15}$$

so that for an incident probe with amplitudes $E_{in}(0) = B_{in}(0)$, taken here for simplicity to be unity, one obtains from Eq. (15)

$$\begin{aligned} r &= \alpha(\varepsilon_0^+ + \varepsilon_{-1}^+) + \beta(\varepsilon_0^- + \varepsilon_{-1}^-) - 1 \\ r &= 1 - \alpha\frac{\kappa_+\varepsilon_0^+ - \kappa_-\varepsilon_{-1}^+}{k_p} - \beta\frac{\kappa_-\varepsilon_0^- - \kappa_+\varepsilon_{-1}^-}{k_p} \\ t &= \alpha(\varepsilon_0^+ + \varepsilon_{-1}^+)e^{iqL} + \beta(\varepsilon_0^- + \varepsilon_{-1}^-)e^{-iqL} \\ t &= \alpha\frac{\kappa_+\varepsilon_0^+ - \kappa_-\varepsilon_{-1}^+}{k_p}e^{iqL} + \beta\frac{\kappa_-\varepsilon_0^- - \kappa_+\varepsilon_{-1}^-}{k_p}e^{-iqL} \end{aligned} \tag{16}$$

where the appropriate expressions for the boundary values of the fields E_p and B_p as obtained from Eq. (14) have been used. The introduction of the ratios

$$X_\pm \equiv \frac{\varepsilon_0^\pm}{\varepsilon_{-1}^\pm} = \frac{-\chi_1}{(1+\chi_0) - (2\kappa_\pm k_c + k_c^2)/k_p^2} \tag{17}$$

and the elimination from Eq. (16) of the two unknowns α and β finally enables one to arrive at the *two-mode approximation* reflection and transmission coefficients

$$\begin{aligned} r &= \frac{2A_-(1+X_+)e^{iqL} - 2B_-(1+X_-)e^{-iqL}}{A_-B_+e^{iqL} - A_+B_-e^{-iqL}} \\ t &= \frac{2A_-(1+X_+) - 2B_-(1+X_-)}{A_-B_+e^{iqL} - A_+B_-e^{-iqL}} \end{aligned} \tag{18}$$

where $A_\pm = (1 \pm \kappa_+/k_p)X_+ + (1 \mp \kappa_-/k_p)$ and $B_\pm = (1 \pm \kappa_-/k_p)X_- + (1 \mp \kappa_+/k_p)$.

We now proceed to examine characteristic photonic structures that may be induced in a sample of ultracold ^{87}Rb atoms. Levels $|1\rangle$ and $|2\rangle$ are here taken to be the hyperfine components $|5S_{1/2}, F=1\rangle$ and $|5S_{1/2}, F=2\rangle$ of the ground $S_{1/2}$ state while $|3\rangle$ is the $|5P_{3/2}, F=1\rangle$ component of the excited $P_{3/2}$ state. For this specific atomic structure we start by examining the Bloch wave vector modes $\kappa_\pm$ near the Brillouin zone band edge for which we show in Fig. 2 the relevant $\mathrm{Im}(\kappa_{\pm}a/\pi)$ and $\mathrm{Re}(\kappa_{\pm}a/\pi - 1)$ around the band edge as obtained from Eq. (11). A photonic bandgap ~0.5 MHz is seen to open up when $\kappa \rightarrow k_c = \pi/a$, due to the off-diagonal term in Eq. (10) proportional to χ_1.

We then study the reflectivity $R = |r|^2$ and the transmitivity $T = |t|^2$ profiles as obtained from Eqs. (18) when Eq. (11) is used. These are plotted for different sample widths in Fig. 3 where over 95% bandgap reflectivity is attained for the wider atomic sample (2.0mm) yet gradually degrading as the sample width decreases. Comparison of Fig.s 3(a) and 3(b) yields, in addition, very low values of absorption ($A = 1 - R - T$) in the bandgap region. This is because the fact that EIT is pretty well established within the whole sample owing to a coupling field that is sufficiently strong even at the SW quasinodes ($\Omega_{c\min} = 5.0MHz$). Moreover, the bandgap width may be increased from ~0.5 MHz to ~3.5 MHz if one uses a twice stronger coupling misaligned at a larger angle (See Fig. 4).

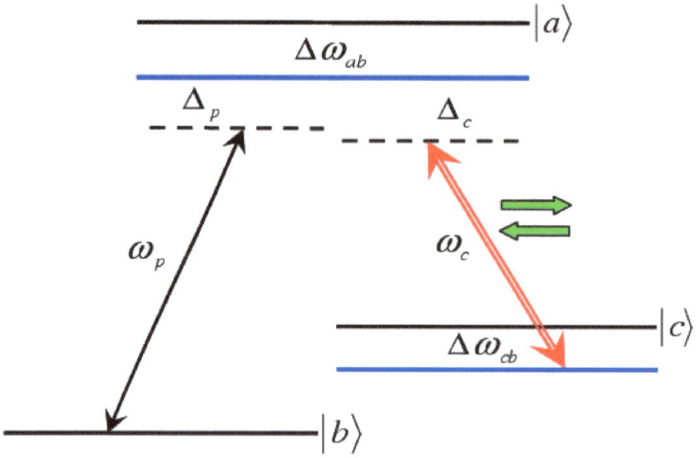

Fig. 5 (Color online) The energy-level configuration for a N-V color center in diamond. The blue lines denote the inhomogeneous line centers of levels $|a\rangle$ and $|c\rangle$ for all active sites in the sample. For negatively-charged N-V centers, levels $|b\rangle$ and $|c\rangle$ belong to the ^{3}A spin triplet manifold with a 2.88 GHz splitting, while level $|a\rangle$ belongs to the strain-split ^{3}E manifold.

The *two-mode approximation* described in this subsection gives a very simple and transparent description of both photonic band structure and linearly responsed spectra in the case of a time-independent *SW* coupling field. The validity of this *two-mode approximation* has been numerically checked through comparison with an exact treatment using the transfer-matrix method as in Ref. [27] (not shown here).

Photonic bandgaps in an inhomogeneous solid driven by a SW light

We start by considering a single three-level impurity in diamond dressed by a coupling field and a probe field as shown in Fig. 5. In the limit of a weak probe, the steady state off-diagonal density matrix element ρ_{ab} [37] is

$$\begin{aligned}
\rho_{ab} &= \frac{i\Omega_p\{[\Gamma_{cb}|X|^2+\gamma_{ab}\Omega_c^2]Y-\Gamma_{cb}\Omega_c^2 X\}}{Z[2\Gamma_{cb}|X|^2+\gamma_{ab}(1+3\Gamma_{cb}/\Gamma_{ab})\Omega_c^2\}}\\
X &= \gamma_{ab}+i(\Delta\omega_{ab}-\Delta\omega_{cb})\\
Y &= \gamma_{cb}+i(\Delta_p+\Delta\omega_{cb})\\
Z &= [\gamma_{ab}+i(\Delta_p-\Delta\omega_{ab})]Y+\Omega_c^2
\end{aligned} \tag{19}$$

with Γ_{ab}, Γ_{ac}, Γ_{cb}, and Γ_{bc} being the population decay rates while γ_{ab}, γ_{ac}, and γ_{cb} denoting the coherence dephasing rates. Moreover, $\Delta\omega_{cb}$ and $\Delta\omega_{ab}$ represent deviations of the resonant frequencies of an impurity from the inhomogeneous line centers contributed by all impurities, Δ_p is the detuning of the probe field from the $|b\rangle \leftrightarrow |a\rangle$ inhomogeneous line center, while Ω_p and Ω_c are Rabi frequencies of the probe and coupling fields, respectively. In deriving Eq. (19), we have assumed that $\Delta_c = 0$ (the coupling is resonant with the $|c\rangle \leftrightarrow |a\rangle$ inhomogeneous line center), $\Gamma_{cb}=\Gamma_{bc}$, $\gamma_{ab}=\gamma_{ac}$, and $\Delta\omega_{ac}=\Delta\omega_{ab}-\Delta\omega_{cb}$. In the case where the coupling has a *SW* pattern generated from the retro-reflection upon a mirror of reflectivity R_m, the resulting squared coupling Rabi frequency varies periodically in the z direction as

$$\Omega_c^2 = \Omega_0^2[(1+\sqrt{R_m})^2\cos^2(k_c z)+(1-\sqrt{R_m})^2\sin^2(k_c z)] \tag{20}$$

with a spatial periodicity $a=\lambda_c/2=\pi c/(\omega_c n_b)$, where $n_b = 2.4$ is the background refractive index of diamond.

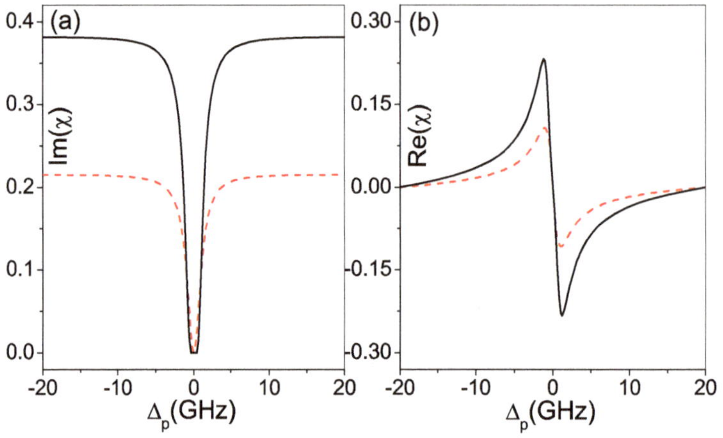

Fig.6 (Color online) Imaginary and real parts of the probe susceptibility for all N-V color centers in diamond driven by a traveling-wave coupling with $\Omega_0 = 20GHz$ and $R_m = 0$. Black-solid and red-dashed curves refer to inhomogeneous Gaussian and Lorentzian profiles, respectively.

The probe linear susceptibility contributed by all active impurities can be easily obtained by performing a proper integration over the inhomogeneous broadening distribution of detunings $\Delta\omega_{cb}$ and $\Delta\omega_{ab}$, which becomes for a Gaussian broadening profile

$$\chi(\Delta_p) = \frac{1}{\pi W_{cb} W_{ab}} \int e^{-\Delta\omega_{cb}^2 / W_{cb}^2} d(\Delta\omega_{cb}) \int e^{-\Delta\omega_{ab}^2 / W_{ab}^2} d(\Delta\omega_{ab}) \frac{N|d_{ab}|^2}{2\varepsilon_0 \hbar} \frac{\rho_{ab}}{\Omega_p} \tag{21}$$

with W_{cb} (W_{ab}) the inhomogeneous linewidths on transitions $|c\rangle \leftrightarrow |b\rangle$ ($|a\rangle \leftrightarrow |b\rangle$). Then we are allowed to evaluate the probe refractive index through $n(\Delta_p) = \sqrt{\chi_b + \chi(\Delta_p)}$ where $\chi_b = n_b^2$. Clearly, the probe refractive index n exhibits periodic behaviors in the z direction so that a photonic bandgap is expected to occur at the Brillouin zone boundary π / a. Using the spatially dependent refractive index, we can further numerically calculate a 2×2 unimodular transfer matrix M [38, 39], which describes the propagation of a monochromatic probe field through a single period of the dressed medium and characterizes the bandgap structure of Bloch wave vectors. In fact, the translational invariance of the periodic medium further requires [27]

$$\begin{pmatrix} E^+(x+a) \\ E^-(x+a) \end{pmatrix} = M \begin{pmatrix} E^+(x) \\ E^-(x) \end{pmatrix} = \begin{pmatrix} e^{i\kappa a}E^+(x) \\ e^{i\kappa a}E^-(x) \end{pmatrix} \tag{22}$$

where E^+ and E^- denote the FD and BD propagating probe electric fields and $\kappa = \kappa' + i\kappa''$ is the probe Bloch wave vector of photonic states. The photonic bandgap structure can be numerically obtained from solutions of the determinantal equation $e^{2i\kappa a} - Tr(M)e^{i\kappa a} + 1 = 0$ with $\det M = 1$ [27]. For a sample of finite length $L = Ka$, with K being the number of standing-wave periods, the total transfer matrix $M_K = M^K$ can be expressed in terms of the single period transfer matrix M, and the reflection and transmission amplitudes for the probe can be written as

$$\begin{aligned} r(\Delta_p) &= \frac{M_{K(12)}(\Delta_p)}{M_{K(22)}(\Delta_p)} \\ t(\Delta_p) &= \frac{1}{M_{K(22)}(\Delta_p)} \end{aligned} \tag{23}$$

from which the reflectivity $R = |r|^2$ and the transmissivity $T = |t|^2$ can be easily calculated.

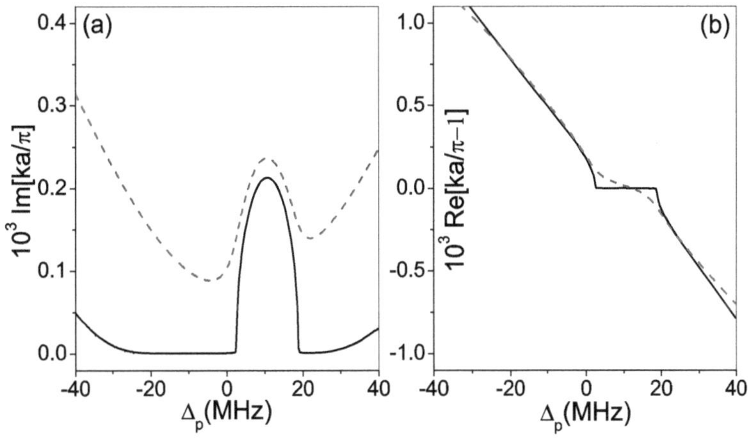

Fig.7 (Color online) Photonic bandgap near the first Brillouin zone boundary induced in a diamond sample with N-V color centers by a standing-wave coupling with $\Omega_0 = 20GHz$, $R_m = 0.617$, and $\theta = 38mrad$. The color notation is the same as in Fig. 6.

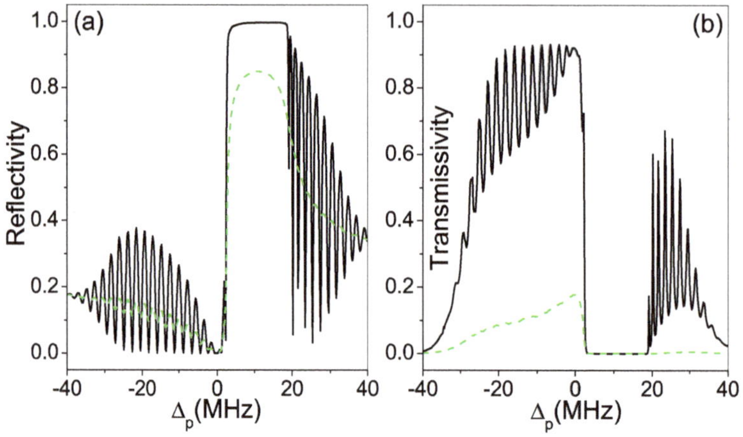

Fig.8 (Color online) Probe reflectivity and transmissivity for a 2.0 mm long diamond sample with N-V color centers. The green-dashed curves correspond to a background absorption of $\mathrm{Im}(\chi_b) = 0.0002$. Other parameters are the same as in Fig. 7.

We adopt here the realistic parameters from a recent experiment on coherent population trapping (CPT) in diamond N-V color centers [40], in particular, using $W_{ab} = 375GHz$ and $W_{cb} = 2.5MHz$ as the values of inhomogeneous linewidths. Probe absorption and dispersion are shown in Fig. 6 where they are further compared with those obtained earlier [28] by using a Lorentzian inhomogeneous broadening profile. Owing to the unphysically long tails of a Lorentzian broadening distribution, the corresponding EIT window turns out to be shallower than the one obtained by using a Gaussian inhomogeneous broadening profile. In the latter case, the reduced values of residual absorption in the EIT region are in turn responsible for the well developed photonic bandgap structure reported in Fig. 7. Correspondingly, probes with frequencies inside the bandgap will experience almost perfect reflection (over 99%) for a sample of finite length as shown in Fig. 8(a). Reflection and transmission in Fig. 8(a) and 8(b) display, in addition, rapid oscillations (fringes) near the bandgap due to the interference arising from the front and back surfaces of the sample. The presence of fringes, however, is tightly related to the background absorption $\mathrm{Im}(\chi_b)$. When this is properly included in the expression of the refractive index used to evaluate the spectra in Fig. 8(a) and 8(b) (green-dashed curves), it can in fact well smooth out the oscillations yet maintaining high values of the reflectivity.

Dynamical optical responses of periodically modulated media

Light propagation dynamics near an induced photonic bandgap

In this subsection, we will investigate the propagation dynamics of an incident probe pulse in ultracold ^{87}Rb atoms around a photonic bandgap induced by a time-independent *SW* coupling. Then it is necessary to resort to the Maxwell wave equations coupled with the density matrix equations, i.e. the coupled *Maxwell-Liouville* equations. In particular, when the probe is very weak, one can set $\rho_{11}=1$ and $\rho_{22}\approx\rho_{33}\approx\rho_{23}\approx 0$ and then reduce Eqs. (2) into

$$\begin{aligned}\partial_t\rho_{21} &= -[\gamma_{12}-i(\Delta_p-\Delta_c)]\rho_{21}+i\Omega_c^*\rho_{31}\\ \partial_t\rho_{31} &= -[\gamma_{13}-i\Delta_p]\rho_{31}+i\Omega_c\rho_{21}+i\Omega_p\end{aligned}\tag{24}$$

with Rabi frequencies $\Omega_c=\Omega_{c+}e^{ik_cz}+\Omega_{c-}e^{-ik_cz}$ and $\Omega_p=\Omega_{p+}e^{ik_cz}+\Omega_{p-}e^{-ik_cz}$. The relevant probe and coupling fields are

$$\begin{aligned}E_p &= (E_{p+}e^{ik_cz}+E_{p-}e^{-ik_cz})e^{-i\omega_pt}\\ E_c &= (E_{c+}e^{ik_cz}+E_{c-}e^{-ik_cz})e^{-i\omega_ct}\end{aligned}\tag{25}$$

with $E_{p\pm}$ and $E_{c\pm}$ being the slowly varying envelopes in time and space. By writing the spin and optical coherence respectively as

$$\begin{aligned}\rho_{21} &= \sum\nolimits_{n=-\infty}^{+\infty}\rho_{21}^{(2n)}e^{i2nk_cz}\\ \rho_{31} &= \sum\nolimits_{n=0}^{\pm\infty}\rho_{31}^{(2n\pm1)}e^{i(2n\pm1)k_cz}\end{aligned}\tag{26}$$

we obtain an infinite set of mutually coupled equations for their (spatial) Fourier components

$$\begin{aligned}\partial_t\rho_{21}^{(2n)} &= -\gamma'_{12}\rho_{21}^{(2n)}+i\Omega_{c+}^*\rho_{31}^{(2n+1)}+i\Omega_{c-}^*\rho_{31}^{(2n-1)}\\ \partial_t\rho_{31}^{(2n\pm1)} &= -\gamma'_{13}\rho_{31}^{(2n\pm1)}+i\Omega_{c+}\rho_{21}^{(2n-1\pm1)}+i\Omega_{c-}\rho_{21}^{(2n+1\pm1)}+i\Omega_{p\pm}\delta_{n,0}\end{aligned}\tag{27}$$

where $\gamma'_{12}=\gamma_{12}-i\Delta_p$ and $\gamma'_{13}=\gamma_{13}-i\Delta_p$ are complex dephasing rates. To examine the propagation dynamics of a FW probe pulse in terms of its Rabi frequencies, the following Maxwell wave equation in the slowly-varying envelope approximation

$$\begin{aligned}\partial_z\Omega_{p+}+\partial_t\Omega_{p+}/c &= +i\Delta k\Omega_{p+}+i\alpha\gamma_{13}\rho_{31}^{(+1)}\\ \partial_z\Omega_{p-}-\partial_t\Omega_{p-}/c &= -i\Delta k\Omega_{p-}-i\alpha\gamma_{13}\rho_{31}^{(-1)}\end{aligned}\tag{28}$$

With $\alpha=N|d_{13}|^2k_p/(2\varepsilon_0\hbar\gamma_{13})$ and $\Delta k=k_p-k_c$ are further derived. And the corresponding boundary conditions are $\Omega_{p+}(z=0,t)=\Omega_{it}(t)$ and $\Omega_{p-}(z=L,t)=0$ with $\Omega_{it}(t)$ being the amplitude of the incident probe pulse.

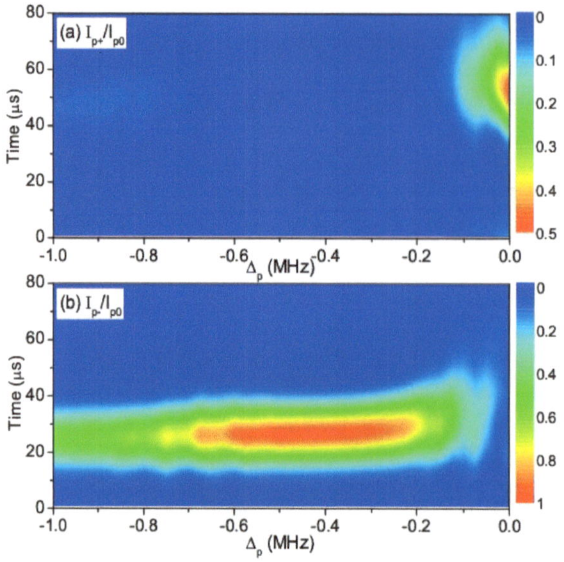

Fig.9 (color online) Reflected components at z=0 (a) and transmitted components at z=L (b) of a weak probe pulse incident upon a L=2.0mm long sample of ultracold ^{87}Rb atoms with $T_0 = 25.0\mu s$ and $\delta T = 4.0\mu s$. Other parameters are the same as in Fig. 2.

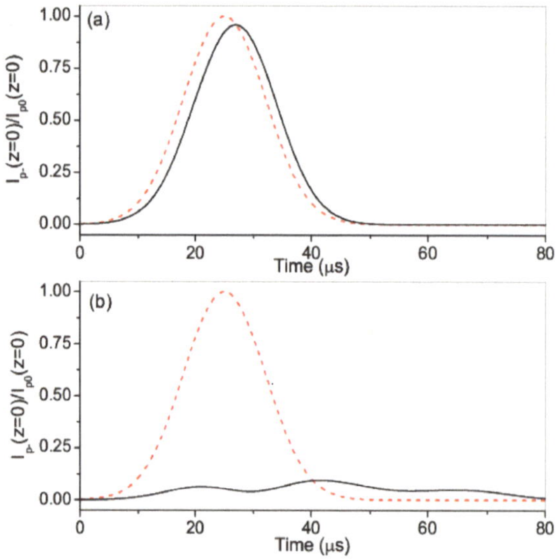

Fig.10 (color online) Reflected probe pulses (black-solid) as seen at z=0. The incident probe pulse (red-dashed) has a carrier frequency of $\Delta_{p0} = -0.4MHz$ in (a) and $\Delta_{p0} = 0$ in (b). Other parameters are the same as in Fig. 9.

The coupled Maxwell-Liouville Eqs. (27-28), when truncated at a suitable value of |n|, enables one to assess with very high precision the propagation dynamics of a probe pulse incident upon the periodic EIT medium induced by a *SW* coupling. In the following, we will assume that $E_{it}(t) = E_0 e^{-(t-T_0)^2/(4\pi\delta T^2)}$ with the central probe detuning denoted by Δ_{p0}. Fig. 9 shows that the reflected pulse at z=0 and the transmitted pulse at z=L are quite sensitive to the central frequency Δ_{p0} of the incident probe. When Δ_{p0} is in the middle of the bandgap, the reflected pulse has little attenuation and distortion compared with the incident probe except a short time delay ($\Delta t = 2.0\mu s$). Conversely, it will experience more and more attenuation and distortion due to the increased transmission and/or absorption as Δ_{p0} moves outward the bandgap. These findings can be further confirmed in Fig. 10 where we compare the reflected pulse profiles with the incident pulse profiles at two specific probe detunings.

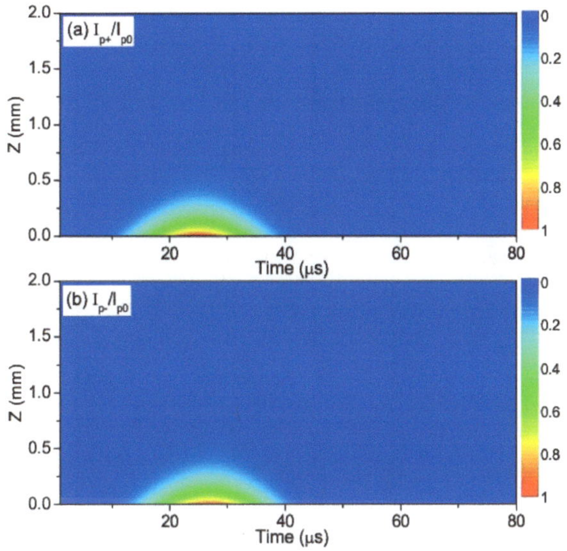

Fig.11 (color online) Scaled intensity distributions of the FW (a) and BW (b) probe pulses inside a sample of ultracold ^{87}Rb atoms with $\Delta_{p0} = -0.4MHz$. Other parameters are the same as in Fig. 9.

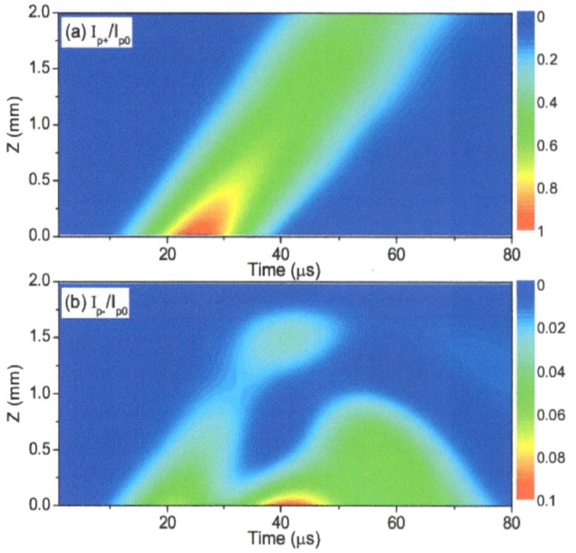

Fig.12 (color online) Scaled intensity distributions of the FW (a) and BW (b) probe pulses inside a sample of ultracold ^{87}Rb atoms with $\Delta_{p0} = 0$. Other parameters are the same as in Fig. 9.

In Fig. 11, we plot the field distributions at different time for the FW and BW probe components with $\Delta_{p0} = -0.4MHz$, and find that they penetrate deeply into the periodic EIT medium even if the reflectivity is nearly perfect. The typical scale of the field penetration is set by the imaginary part of the optical Bloch wavevector within the photonic bandgap. If the penetration length is defined as $L_p = 1/[2\,\mathrm{Im}(\kappa)]$, one obtains $L_p = 0.176mm$ for $\Delta_p = -0.4MHz$ from Fig. 2(a), which is consistent with Fig. 11 where $I_{p+}(z = 0.189mm)/I_{p+}(z = 0) = 1/e$ at $T = 40\mu s$. In Fig. 12, with another central frequency $\Delta_{p0} = 0$, we plot once again the field distributions for the FW and BW probe components. We can see that, due to the imperfect reflectivity, both FW and BW components can reach the end of the atomic sample, and their distributions change dramatically with time due to the interference resulting from the multiple feedback at the boundaries.

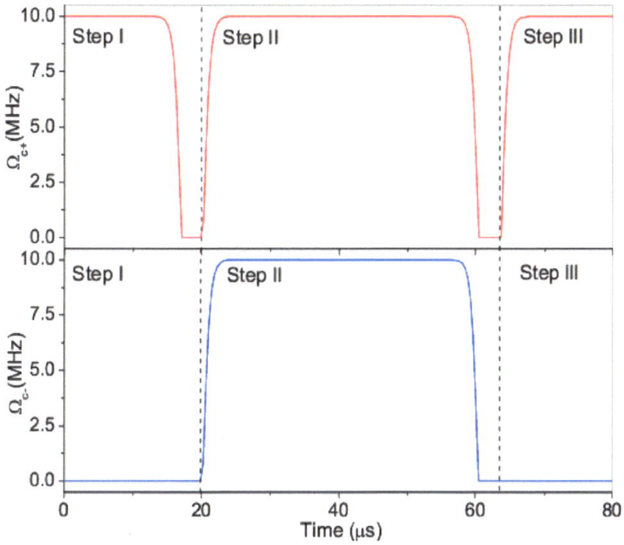

Fig. (13) (color online) Time modulation of the FW and BW components of a SW coupling.

Stationary light generation using a perfect SW grating

In this subsection, we will study the propagation dynamics of a probe pulse incident upon a sample of ultracold ^{87}Rb atoms dressed by a time-modulated SW coupling, and pay special attention to the stationary light generation during the process where the two coupling components are switched on and off as in Fig. 13. In Fig. 14, we focus on Ω_{p+} and Ω_{p-} together with several spin and optical coherences $\rho_{21}^{(0)}$, $\rho_{31}^{(+1)}$, $\rho_{21}^{(+10)}$, and $\rho_{31}^{(+11)}$ using Eqs. (27) truncated at |n|=40 and Eqs. (28). The whole dynamic process of stationary light generation is described in terms of three successive steps regarding to the modulation of Ω_{c+} and Ω_{c-}.

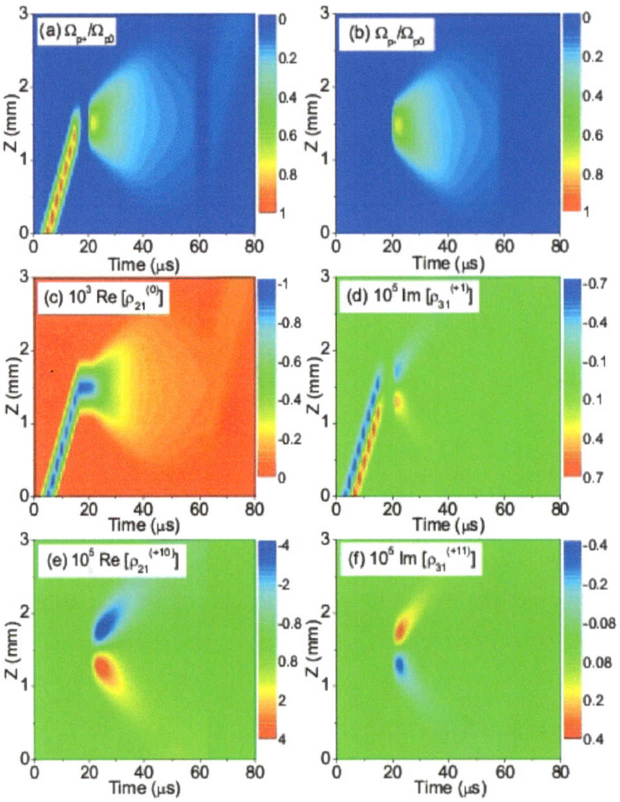

Fig.14 (color online) Nonlinear propagation dynamics of an incident probe pulse having a 1/e full width of $4.0\mu s$ with the scaled Ω_{p+}, Ω_{p-}, $\rho_{21}^{(0)}$, $\rho_{31}^{(+1)}$, $\rho_{21}^{(+10)}$, and $\rho_{31}^{(+11)}$ illustrated. The probe and coupling detunings are set as $\Delta_p = \Delta_c = 0$. The relevant atomic parameters are $\gamma_{12} = 1.0kHz$, $\gamma_{13} = 6.0MHz$, $\lambda_p = 780nm$, $N = 1.0\times10^{13}cm^{-3}$, and $d_{13} = 1.465\times10^{-29}Cm$, respectively. Eqs. (27) are truncated at |n|= 40.

In the first step, in the presence of a FW coupling, only Ω_{p+}, $\rho_{21}^{(0)}$, and $\rho_{31}^{(+1)}$ have nonzero values and the optical coherence $\rho_{31}^{(+1)}$ is about two orders smaller than the spin coherence $\rho_{21}^{(0)}$. When the FW coupling is switched off, the FW probe traveling at a very slow velocity decreases to zero together with the optical coherence $\rho_{31}^{(+1)}$. In the meantime, all characteristics of the FW probe are mapped onto the stationary spin coherence $\rho_{21}^{(0)}$ at the sample center.

In the second step, both FW and BW probes are generated and frozen in the sample together with an absorptive and dispersive (atomic) grating when a perfect *SW* coupling is switched on. This atomic grating is composed of all spin and optical coherences involved in the truncated Eqs. (27), which are nonzero as shown by a few examples in Fig. 14. The two probes are mapped onto $\rho_{21}^{(0)}$ once again when the *SW* coupling is switched off. Simultaneously, the atomic grating becomes non-absorptive and non-dispersive because it is only composed of spin coherences, which don't decay through spontaneous emission. Moreover, the FW and BW probes first quickly increase in amplitude as the *SW* coupling is being switched on and then experience dramatic loss and diffusion after the *SW* coupling becomes constant.

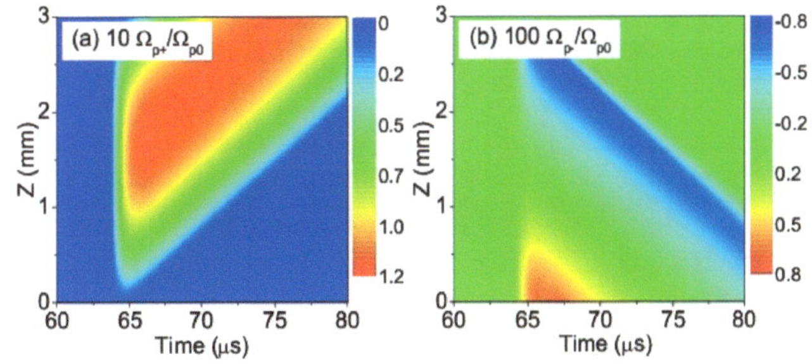

Fig. (15) Blowing-up of the second two plots in Fig. 14 during the last $20\mu s$.

In the third step, the FW and BW probes are retrieved in an asymmetric way by switching on a FW coupling, which is accompanied by the death of the atomic grating composed of only spin coherences. It is easy to see that, if we simply set $\Omega_{c-} = 0$, Eqs. (27) will become decoupled in the following way

$$\begin{aligned}
\partial_t \rho_{21}^{(0)} &= -\gamma_{12}\rho_{21}^{(0)} + i\Omega_{c+}^{*}\rho_{31}^{(+1)} \\
\partial_t \rho_{31}^{(+1)} &= -\gamma_{13}\rho_{31}^{(+1)} + i\Omega_{c+}\rho_{21}^{(0)} + i\Omega_{p+} \\
\partial_z \Omega_{p+} &= -\partial_t \Omega_{p+}/c + i\alpha\gamma_{13}\rho_{31}^{(+1)}
\end{aligned} \tag{29}$$

$$\begin{aligned}
\partial_t \rho_{21}^{(-2)} &= -\gamma_{12}\rho_{21}^{(-2)} + i\Omega_{c+}^{*}\rho_{31}^{(-1)} \\
\partial_t \rho_{31}^{(-1)} &= -\gamma_{13}\rho_{31}^{(-1)} + i\Omega_{c+}\rho_{21}^{(-2)} + i\Omega_{p-} \\
\partial_z \Omega_{p-} &= +\partial_t \Omega_{p-}/c - i\alpha\gamma_{13}\rho_{31}^{(-1)}
\end{aligned} \tag{30}$$

$$\begin{aligned}
\partial_t \rho_{21}^{(+2)} &= -\gamma_{12}\rho_{21}^{(+2)} + i\Omega_{c+}^{*}\rho_{31}^{(+3)} \\
\partial_t \rho_{31}^{(+3)} &= -\gamma_{13}\rho_{31}^{(+3)} + i\Omega_{c+}\rho_{21}^{(+2)}
\end{aligned} \tag{31}$$

$$\begin{aligned}
\partial_t \rho_{21}^{(-4)} &= -\gamma_{12}\rho_{21}^{(-4)} + i\Omega_{c+}^{*}\rho_{31}^{(-3)} \\
\partial_t \rho_{31}^{(-3)} &= -\gamma_{13}\rho_{31}^{(-3)} + i\Omega_{c+}\rho_{21}^{(-4)}
\end{aligned} \tag{32}$$

where only the first four typical sets are listed. Then, in a short time of the order of $T \approx 1/\gamma_{31}$, except $\rho_{21}^{(0)}$ coupled with $\rho_{31}^{(+1)}$ and $\rho_{21}^{(-2)}$ coupled with $\rho_{31}^{(-1)}$ in the EIT regime, all other spin coherences will quickly decay to zero after transferring into their neighboring optical coherences. In this case, only Eqs. (11) and (12) have to be reserved, which then result in the generation of decoupled FW and BW probes. The two probes are very different in amplitude and in profile as $\rho_{21}^{(0)}$ and $\rho_{21}^{(-2)}$ do, which is shown in Fig. 15 by blowing up the first two plots in Fig. 14 during the last $20\mu s$.

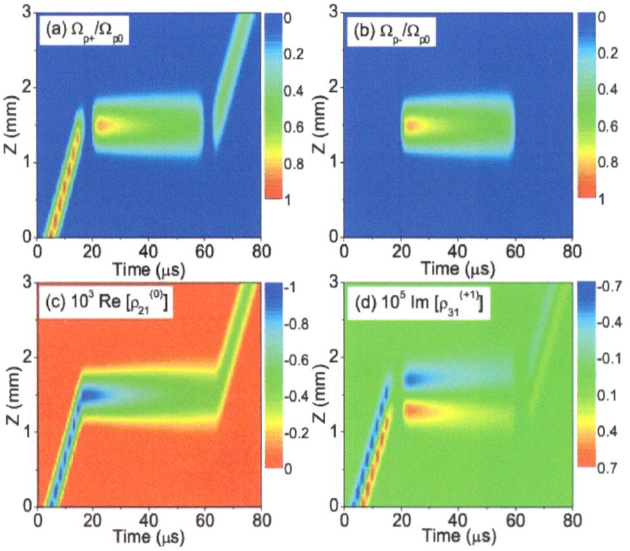

Fig.16 (color online) Nonlinear propagation dynamics of the same incident probe pulse and the accompanied coherence components with the same parameters as in Fig. 14 except that Eqs. (27) are truncated at |n|=0.

By truncating Eq.s (27) at |n|=0, we simulate in Fig. 16 once again the nonlinear propagation and generation dynamics of Ω_{p+}, Ω_{p-}, $\rho_{21}^{(0)}$, and $\rho_{21}^{(+1)}$ with the same parameters as in Fig. 14. It is clear that the FW and BW probes experience little loss and diffusion due to the lack of higher-order spin and optical coherences when the *SW* coupling is switched on. To better understand the difference between Fig. 14 and Fig. 16, we plot in Fig. 17 a diagram for the mutual coupling between the spin and optical coherences starting from $\rho_{21}^{(0)}$, which can be directly inferred from Eqs. (27). It is clear that, to generate the FW and BW probes, $\rho_{31}^{(\pm1)}$ should not be exactly zero although they may be very small. In fact, $\rho_{31}^{(\pm1)}$ are the unique bridges between the source $\rho_{21}^{(0)}$ and the outcome $\Omega_{p\pm}$ during the retrieval process where characteristics of Ω_{p+}

mapped on $\rho_{21}^{(0)}$ are transferred back to both Ω_{p+} and Ω_{p-}. If $\rho_{31}^{(\pm 1)}$ is nonzero, then all

spin and optical coherences involved in the truncated Eqs. (27) will be successively generated. So

it is the excitation of these higher-order spin and optical coherences (i.e. the absorptive and

dispersive grating) that results in the dramatic probe loss and diffusion shown in Fig. 14. It is also

clear that we can not make an adiabatic assumption to ignore all the excited spin and optical

coherences when the coupling is in the SW configuration.

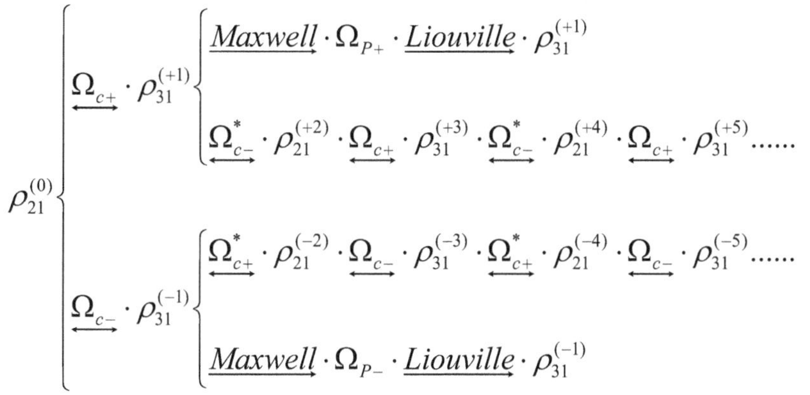

$$\rho_{21}^{(0)} \begin{cases} \underleftrightarrow{\Omega_{c+}} \cdot \rho_{31}^{(+1)} \begin{cases} \underrightarrow{Maxwell} \cdot \Omega_{P+} \cdot \underrightarrow{Liouville} \cdot \rho_{31}^{(+1)} \\\\ \underleftrightarrow{\Omega_{c-}^*} \cdot \rho_{21}^{(+2)} \cdot \underleftrightarrow{\Omega_{c+}} \cdot \rho_{31}^{(+3)} \cdot \underleftrightarrow{\Omega_{c-}^*} \cdot \rho_{21}^{(+4)} \cdot \underleftrightarrow{\Omega_{c+}} \cdot \rho_{31}^{(+5)} \cdots\cdots \end{cases} \\\\ \underleftrightarrow{\Omega_{c-}} \cdot \rho_{31}^{(-1)} \begin{cases} \underleftrightarrow{\Omega_{c+}^*} \cdot \rho_{21}^{(-2)} \cdot \underleftrightarrow{\Omega_{c-}} \cdot \rho_{31}^{(-3)} \cdot \underleftrightarrow{\Omega_{c+}^*} \cdot \rho_{21}^{(-4)} \cdot \underleftrightarrow{\Omega_{c-}} \cdot \rho_{31}^{(-5)} \cdots\cdots \\\\ \underrightarrow{Maxwell} \cdot \Omega_{P-} \cdot \underrightarrow{Liouville} \cdot \rho_{31}^{(-1)} \end{cases} \end{cases}$$

Fig. 17: Schematic diagram of the successive generation and mutual coupling of the FW and BW

probes as well as the higher-order spin and optical coherences starting from the initially prepared

zero-order spin coherence $\rho_{21}^{(0)}$.

To see whether the truncation at |n|=40 is suitable, we further plot in Fig. 18 the dynamical

evolution of Ω_{p+}, Ω_{p-}, $\rho_{21}^{(0)}$, $\rho_{31}^{(+1)}, \rho_{21}^{(+10)}$, and $\rho_{31}^{(+11)}$ with |n|=30. The difference between

the curves for |n|=40 and those for |n|=30 is very small, so we may conclude that the truncation at

|n|=40 is enough to provide correct results with very high precision for this sample of ultracold

atoms. Thus one cannot obtain a good stationary light within an ultracold atomic sample dressed

by a perfect *SW* coupling field in the Lambda configuration. To have a stationary light with little

loss and diffusion, one feasible way is to use a dichromatic coupling field with two components of

different carrier frequencies and opposite traveling directions [21]. When the frequency difference

between the two coupling components is large enough, all higher-order spin and optical

coherences are well suppressed due to their gradually increasing multi-photon detunings, which surely facilitate the formation of a good stationary light. A double Lambda system with two coupling fields traveling in the opposite directions may also be adopted to generate a good stationary light signal [21, 41].

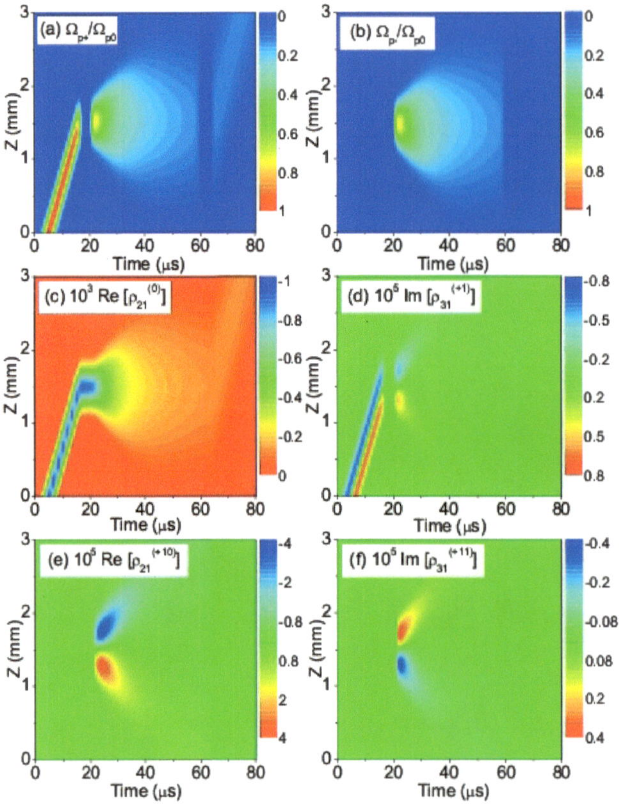

Fig.18 (color online) Nonlinear propagation dynamics of the same incident probe pulse and the accompanied coherence components with the same parameters as in Fig. 2 except that Eqs. (27) are truncated at |n|=30.

Conclusions

To summarize, we have briefly reviewed a few recent works on the steady and dynamical optical responses of ultracold atoms and diamond N-V color centers dressed by a *SW* coupling in the Lambda configuration.

We first investigate the steady optical responses, i.e. the photonic bandgap structures and the reflection and transmission spectra, of our considered media to a cw probe field with the two-mode approximation method and the transfer-matrix method. Starting from the dressed susceptibility in the weak probe limit, we find that a photonic bandgap of 0.5 MHz in width can be dynamically induced by an imperfect *SW* coupling in ultracold ^{87}Rb atoms and the probe reflectivity in this bandgap may homogeneously exceed 95%. As extended to the diamond N-V color centers with a Gaussian profile of inhomogeneous broadening, the dynamically induced photonic bandgap is found to have a width of 20 MHz and the probe reflectivity is found to homogeneously exceed 99% due to the much better suppressed absorption.

Then we consider the dynamic optical responses of ultracold ^{87}Rb atoms with the spatially expanded Maxwell-Liouville equations in two different situations. In the first situation with an imperfect *SW* coupling, the probe pulse can be either perfectly reflected with a shorter time delay or partially transmitted with a longer time delay depending on whether its carrier frequencies are contained in the photonic bandgap or not. Moreover, the probe pulse is found to penetrate into the media with complicated field distributions due to the interference between the FW and BW components. In the second situation with a perfect *SW* coupling, we find that a stationary light can be generated from the zero-order atomic spin coherence. The stationary light signal diffuses and decays quickly when the Maxwell-Liouville equations are properly truncated at $|n|=40$ to collect most important higher-order spin and optical coherences, but experiences little loss and diffusion when the Maxwell-Liouville equations are arbitrarily truncated at $|n|=0$ to unphysically neglect the higher-order spin and optical coherences.

Finally we conclude that the higher-order spin and optical coherences facilitate the formation of a good photonic bandgap, but hamper the generation of a good stationary light. To have a stationary light with little loss and diffusion, one should try to suppress the higher-order spin and optical coherences, e.g. by using a dichromatic coupling field in the single Lambda system or by adopting the double Lambda system driven by two different coupling fields [21, 41].

Acknowledgement

This work is supported by the National Key Basic Research Program of China (2006CB921103), the Program for New Century Excellent Talents in the Universities of China (NCET-06-0309), the Program for Distinguished Young Scholar in Jilin Province of China (20070121), the National Natural Science Foundation of China (10874057), the PRIN 2006-021037 grant and the Azione

Integrata IT09L244H5 of MIUR Italy, and the CNR contract 0008233 Italy.

References

1. Boller KJ, Imamoglu A, Harris SE. Observation of electromagnetically induced transparency. Physical Review Letters 1991 May 20;66(20):2593-2596.
2. Xiao M, Li YQ, Jin SZ, Gea-Banacloche J. Measurement of dispersive properties of electromagnetically induced transparency in rubidium atoms. Physical Review Letters 1995 Jan 30;74(5): 666-669.
3. Harris SE. Electromagnetically induced transparency. Physics Today 1997 Jul;50 (7):36-42.
4. Fleischhauer M, Imamoglu A, Marangos JP. Electromagnetically induced transparency: Optics in coherent media. Reviews of Modern Physics 2005 Jul 12;77(2):633-673.
5. Hau LV, Harris SE, Dutton Z et al. Light speed reduction to 17 meters per second in an ultracold atomic gas. Nature (London) 1999 Feb 18;397:594-598.
6. Turukhin AV, Sudarshanam VS, Shahriar MS et al. Observation of ultraslow and stored light pulses in a solid. Physical Review Letters 2001 Dec 20; 88(2):023602(4).
7. Fleischhauer M, Lukin MD. Dark-state polaritons in electromagnetically induced transparency. Physical Review Letters 2000 May 29;84(22):5094-5097.
8. Liu C, Dutton Z, Behroozi CH et al. Observation of coherent optical information storage in an atomic medium using halted light pulses. Nature (London) 2001 Jan 25;409:490-493.
9. Longdell JJ, Fraval E, Sellars MJ et al. Stopped light with storage times greater than one second using electromagnetically induced transparency in a solid. Physical Review Letters 2005 Aug 2;95(6):063601(4).
10. Chaneliere T, Matsukevich DN, Jenkins SD et al. Storage and retrieval of single photons transmitted between remote quantum memories. Nature (London) 2005 Dec 8;438:833-836.
11. Choi KS, Deng H, Laurat J et al. Mapping photonic entanglement into and out of a quantum memory. Nature (London) 2008 Mar 6; 452: 67-71.
12. Appel J, Figueroa E, Korystov D et al. Quantum memory for squeezed light. Physical Review Letters 2008 Mar 5;100(9):093602(4).
13. Harris SE, Hau LV. Nonlinear optics at low light levels. Physical Review Letters 1999 Jun 7; 82(23): 4611-4614.
14. Braje DA, Balic V, Yin GY et al. Low-light-level nonlinear optics with slow light. Physical Revew A 2003 Oct 15;68(4):041801(4).
15. Matsko AB, Novikova A, Welch GR et al. Enhancement of Kerr nonlinearity by multiphoton coherence. Optics Letters 2003 Jan 15;28(2):96-98.
16. Ottaviani C, Vitali D, Artoni M and et al. Polarization qubit phase gate in driven atomic media. Physical Review Letters 2003 May 15;90(19):197902(4).
17. Rebic S, Vitali D, Ottaviani C and et al. Polarization phase gate with a tripod atomic system. Physical Revew A 2004 Sep 21;70(3):032317(8).
18. Harris SE and Yamamoto Y. Photon switching by quantum interference. Physical Review Letters 1998 Oct 26;81(17):3611-3614.
19. Wu JH, Gao JY, Xu JH et al. Ultrafast all optical switching via tunable Fano interference. Physical Review Letters 2005 Jul 25;95(5):057401(4).
20. Bajcsy M, Zibrov AS, Lukin MD, Stationary pulses of light in an atomic medium. Nature

(London) 2003 Dec 11;426:638-641.
21. Lin YW, Liao WT, Peters T et al. Stationary light pulses in cold atomic media and without Bragg gratings. Physical Review Letters 2009 May 28;102(21):213601(4).
22. Andre A, Bajcsy M, Zibrov AS and et al. Nonlinear optics with stationary pulses of light. Physical Review Letters 2005 Feb 15;94(6):063902(4).
23. Ham BS. Spatiotemporal quantum manipulation of traveling light: Quantum transport. Applied Physics Letters 2006 Mar 23;88:121117(3).
24. Zimmer FE, Andre A, Lukin MD and et al. Coherent control of stationary light pulses. Optics Communications 2006 Aug 15;264(2):441-453.
25. Hansen KR, Molmer K. Trapping of light pulses in ensembles of stationary lambda atoms. Physical Revew A 2007 May 2;75(5):053802(8).
26. Kuang SQ, Wan RG, Du P et al. Transmission and reflection of electromagnetically induced absorption grating in homogeneous atomic media. Optics Express 2008 Sep 29;16(20): 15455-15462.
27. Artoni M, La Rocca GC. Optically tunable photonic stop bands in homogeneous absorbing media. Physical Review Letters 2006 Feb 24;96(7):073905(4).
28. He QY, Xue Y, Artoni M et al. Coherently induced stop-bands in resonantly absorbing and inhomogeneously broadened doped crystals. Physical Revew B 2006 May 30;73(9):195124(7).
29. Wu JH, La Rocca GC, Artoni M. Controlled light-pulse propagation in driven color centers in diamond. Physical Revew B 2008 Mar 26;77(11):113106(4).
30. Wu JH, Artoni M, La Rocca GC. Controlling the photonic band structure of optically driven cold atoms. Journal of the Optical Society of America B 2008 Nov;25(11):1840-1849.
31. Andre A and Lukin MD. Manipulating light pulses via dynamically controlled photonic band gap. Physical Review Letters 2002 Sep 16;89(14):143602(4).
32. Su XM and Ham BS. Dynamic control of the photonic band gap using quantum coherence. Physical Revew A 2005 Jan 31;71(1):013821(5).
33. Prineas JP, Zhou JY, Kuhl J et al. Ultrafast ac Stark effect switching of the active photonic band gap from Bragg-periodic semiconductor quantum wells. Applied Physics Letters 2002 Dec 2; 81(23):4332-4334.
34. Sadeghi Sm, Li W, van Driel HM. Coherently induced one-dimensional photonic band gap. Physical Revew B 2004 Feb 13;69(7):073304(4).
35. Sakoda K. Optical properties of photonic crystals. Berlin: Springer; 2001.
36. Riley K, Hobson, Bence S. Mathematical Methods for Physics and Engineering, 3rd edition. Cambridge: Cambridge University Press; 2006.
37. Kuznetsova E, Kocharovskaya O, Hemmer PR et al. Atomic interference phenomena in solids with a long-lived spin coherence. Physical Revew A 2002 Dec 6;66(6):063802(13).
38. Born M, Wolf E. Principles of Optics, 7th (expanded) edition. Cambridge: Cambridge University Press; 1999.
39. Artoni M, La Rocca GC, Bassani F. Resonantly absorbing one-dimensional photonic crystals. Physical Revew E 2005 Oct 10;72(4):046604(11).
40. Santori C, Fattal D, Spillane SM et al. Coherent population trapping in diamond N-V centers at zero magnetic field. Optics Express 2006 Aug 21;14(17):7986-7993.
41. Zimmer FE, Otterbach J, Unanyan RG et al. Dark-state polaritons for multicomponent and stationary light fields. Physical Revew A 2008 Jun 16;77(6):063823(6).

Chapter 6. Atomic Coherence and Optical Storage

Hai-Hua Wang, Ai-Jun Li, and Jin-Yue Gao
Jilin University

Abstract: Atomic coherence is the interaction process between light and atoms, that is one or more coherent fields couple the different atomic states and cause the quantum interference between the different transition channels. Atomic coherence has led to many interesting and unexpected consequences, such as the Hanle effects, electromagnetically induced transparency (EIT), coherence population trapping (CPT), stimulated Raman adiabatic passage (STIRAP), spontaneous emission control, resonant enhancement of optical nonlinearity, slow and superluminal light propagation, quantum light storage, etc. In which, light storage based on atomic coherence plays an important role in the coherent control of light pulse information. In this chapter, we present some previous works on atomic coherence and optical storage. Firstly, we present the storage and recovery of light pulse based on F-STIRAP, which is fundamentally different from the conventional EIT-based process. Secondly, we present the applications of light storage based on EIT in a $Pr^{3+}:Y_2SiO_5$ crystal, which includes the erasure of stored optical information, all-optical routing by light storage, and the coherent control of double light pulses. At last, we propose theoretically and demonstrate experimentally a method to control the atomic coherence by a STIRAP or F-STIRAP process.

Introduction

Light is an excellent carrier of information because it possesses the fastest propagation speed and provides large communication bandwidth, but it is also difficult to localize and store. So the ability to manipulate the behavior of light becomes very important in classical and quantum optics. Recently, the storage and release of light pulse in atomic ensemble by using the effect of electromagnetically induced transparency (EIT) [1-2] have been proposed theoretically [3] and demonstrated experimentally [4,5]. This light-storage technique is based on the quantum state transfer between light and matter. In a three-level lambda-type EIT system, a weak probe pulse can be completely halted in the atoms by switching off the control field, and then the probe pulse is stored in the created atomic spin coherence, and subsequently released by the reverse process. This light-storage technique provides a new way to manipulate the behavior of light, and attracted much attention. Subsequently, many interesting studies on light storage are reported [6-17].

Light storage can be realized under different mechanisms. It is well known that the light information is stored in the created spin coherence in the process of light storage. Different buildup mechanisms of spin coherence correspond to different light-storage mechanisms. Light storage based on dynamic EIT [4,5] and Coherent population trapping (CPT) [6] has been reported, respectively. Fractional stimulated Raman adiabatic passage (F-STIRAP) [18-22] is usually used to create maximal spin coherence. During the building of spin coherence by F-STIRAP, the medium also memorizes the pulse information. The pulse information exists as spin coherence in the medium. So the creation of spin coherence by F-STIRAP also involves the process of light information storage [23,24]. This is a different mechanism from the conventional EIT-based processes, and it is very useful in quantum information and all-optical communication.

Many experimental studies on light storage have been carried out in atomic gases. However, the atomic motion in atomic gases limits the applications of light storage. Solid-state materials are considered as promising candidates for corresponding experimental demonstrations. The obvious advantages of solids are high density of atoms, compactness, and absence of atomic diffusion. However, most solid materials have relatively broad optical linewidths and fast decoherence rate, which limit the achievable light storage and atomic coherence. Now it is found that $Pr^{3+}:Y_2SiO_5$ (Pr:YSO) has sharp spectrum structure and long spin coherence time, which is suitable for the experimental demonstration of atomic coherence [25,26]. Recently, EIT [27,28], enhanced four-wave mixing [29-31], quantum switching [32], all-optical routing [33,34], light storage [35-39], and stimulated Raman adiabatic passages (STIRAP) [18,20] have been experimentally reports in solids.

Most reports related to atomic coherence concentrate on the way to generate or employ atomic coherence, such as creation of coherent superposition states, quantum logical gates, slow light storage, etc. A further interesting topic related to the atomic coherence is how to manipulate the existed coherence. Recently, a method is developed to control the atomic coherence in theory and experiment [40]. By a STIRAP or F-STIRAP process, the atomic coherence can be completely transferred or arbitrarily contributed among the different ground-state levels. This technique can be applied to the information conversion in slow light storage, quantum logical gates and so on.

In this chapter, we will give a brief review of our recent research work on light storage and atomic coherence. We firstly reported the storage and recovery of light signal based on F-STIRAP, then reported the applications of light storage based on EIT in Pr:YSO crystal. At last, we reported the control of atomic coherence by a STIRAP or F-STIRAP.

Storage and release of light pulse based on F-STIRAP

F-STIRAP is usually used to create maximal spin coherence. During the building of spin coherence by F-STIRAP, the medium memorizes the pulse information. The pulse information exists as spin coherence in the medium. So F-STIRAP is used as a technique of light storage. This technique is a different mechanism from the conventional EIT-based processes [23,24]. In this section, we present our experimental research on light storage and release based on F-STIRAP in atomic vapor and Pr:YSO crystal.

A. Storage and release of light pulse based on F-STIRAP in atomic vapor

An experiment of light storage via maximum coherence prepared by F-STIRAP is reported in a three-level lambda-type system of ^{87}Rb atoms. This work is different from the conventional light storage experiment. Instead of EIT, the maximum coherence is used for optical storage. The write control laser pulse $(\lambda_1 = 794.9842nm)$ is turned on to store the probe pulse

$(\lambda_2 = 794.9698nm)$ in collective atomic medium. The two pulses with the same back edge drive the atoms of ensemble into a maximum coherent superposition between the lower levels. After a time interval, we turn on the retrieve control pulse at 794.9842nm, the recovered pulse at 794.9698nm is released. Similarly, we turn on the retrieve control pulse at 794.9698nm, the revived pulse at 794.9842nm is released.

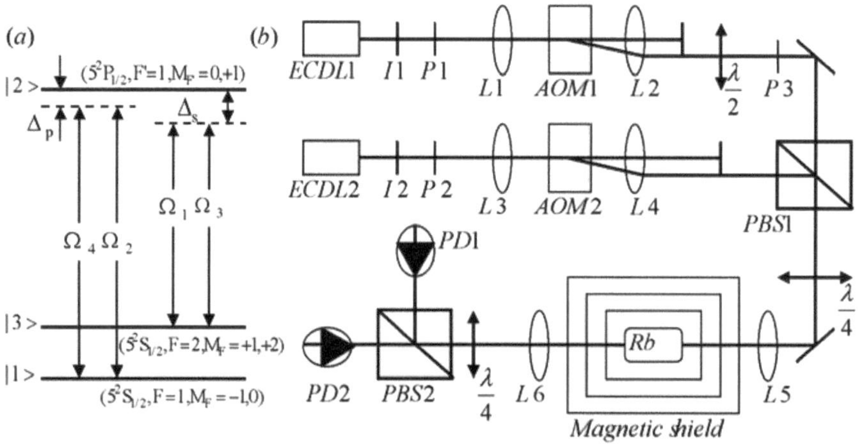

Fig. 1. (a) Energy-level diagram of the lambda-type system in [87]Rb. (b) Schematic of the experimental setup.

A three-level lambda-type system is shown in Fig. 1(a). Hyperfine levels $\left|5^2S_{1/2}, F=1, M_F=-1,0\right\rangle$ and levels $\left|5^2S_{1/2}, F=2, M_F=+1,+2\right\rangle$ are used as the two ground states, labeled as $\left|1\right\rangle$ and $\left|3\right\rangle$, respectively. The upper state is $\left|5^2P_{1/2}, F'=1, M_F=0,+1\right\rangle$, showed as level $\left|2\right\rangle$. The write control pulse with Rabi frequency Ω_1 $(\lambda_1 = 794.9842nm)$ couples $\left|3\right\rangle$ and $\left|2\right\rangle$ with a frequency detuning of Δ_s. It is right circularly polarized. The power of the write control pulse is 6mW, with duration of 170ns, which can drive all populations into state $\left|1\right\rangle$ before the probe pulse is applied. The probe pulse with Rabi frequency Ω_2 $(\lambda_2 = 794.9698nm)$ couples $\left|1\right\rangle$ and $\left|2\right\rangle$ with a frequency detuning of Δ_p. It is left circularly polarized. The power of the probe pulse is 6.7mW, with duration of 30ns. The two pulses prepare the Rb atom with maximal atomic coherence defined as the situation that density matrix elements satisfy $\rho_{11} \approx \rho_{33} \approx |\rho_{13}| \approx 0.5$. The retrieve control laser turns to the $\left|3\right\rangle \rightarrow \left|2\right\rangle$ transition with Rabi frequency Ω_3 $(\lambda_1 = 794.9842nm)$ and reads out the restored optical pulse with Rabi frequency Ω_4 $(\lambda_2 = 794.9698nm)$. The time delay between the end of the write control pulse and the beginning of the retrieve control pulse is 80ns. The duration of the retrieve control pulse is 30ns. Similarly, the retrieve control laser can turn to the $|1\rangle \to |2\rangle$ transiton with Rabi frequency Ω_4 and obtain the released pulse with Rabi frequency Ω_3. The interaction Hamiltonian in the rotating wave and dipole moment approximations for the three-level system is

$$H^I = \hbar\Delta_p|2\rangle\langle 2| + \hbar(\Delta_p - \Delta_s)|3\rangle\langle 3| - \frac{\hbar}{2}[(\Omega_2 + \Omega_4)|2\rangle\langle 1| + (\Omega_1 + \Omega_3)|2\rangle\langle 3| + H.C.] \quad (1)$$

Here, $\Omega_i = |\mu_i|E_i/\hbar$ $(i = 1,2,3,4)$ is the Rabi frequency of the respective fields, μ_i is the electrical dipole matrix element, E_i is the amplitude of fields, $\Delta_p = \omega_{21} - \omega_2$ and $\Delta_s = \omega_{23} - \omega_1$ are the laser detunings from the atomic resonance, and ω_1 and ω_2 are the frequencies of fields. The elements of the density matrix are given by the Liouville equation

$$\rho_{ij} = -\frac{i}{\hbar}\sum(H^I_{ik}\rho_{kj} - \rho_{ik}H^I_{kj}) - \frac{1}{2}\sum(\Gamma_{ik}\rho_{kj} + \rho_{ik}\Gamma_{kj}), \quad (2)$$

Where $\Gamma_{ik}\,(i,k = 1,2,3)$ is the spontaneous emission decay rate from the state $|i\rangle$ to the state $|k\rangle$. The equations for the fields are

$$\frac{\partial(\Omega_1 + \Omega_3)}{\partial\xi} = i\eta_{23}\rho_{32}, \quad (3)$$

$$\frac{\partial(\Omega_2 + \Omega_4)}{\partial\xi} = i\eta_{21}\rho_{12}, \quad (4)$$

Where $\eta_{23} = \omega_1 N|\mu_{23}|^2/\varepsilon_0 c\hbar$ and $\eta_{21} = \omega_2 N|\mu_{21}|^2/\varepsilon_0 c\hbar$ are the coupling constants, N is the density of medium, ε_0 is the permittivity of the vacuum, and c is the speed of light in vacuum.

The experimental setup is shown in Fig. 1(b). Both ECDL1 and ECDL2 are the DL-100 external-cavity diode lasers with linearly polarized output beams. One beam passes through an acousto-optic modulator that is used to switch on and off the laser to generate pulse as the write control pulse. The polarization of the write control pulse is rotated by 90^0 after passing through $\lambda/2$ wave plate. The other beam passes through the other acousto-optic modulator that is used to turn on and off the laser to generate the probe pulse. The focus length of L1, L2, L3, and L4 is 5cm. The two beams are combined with a polarizing beam splitter. Then they are changed to opposite circularly polarized beams with a $\lambda/4$ wave plate. We perform the experiment in 3.5cm long and 2.5cm diameter glass cell containing atomic Rb vapor at $85-98^0C$,

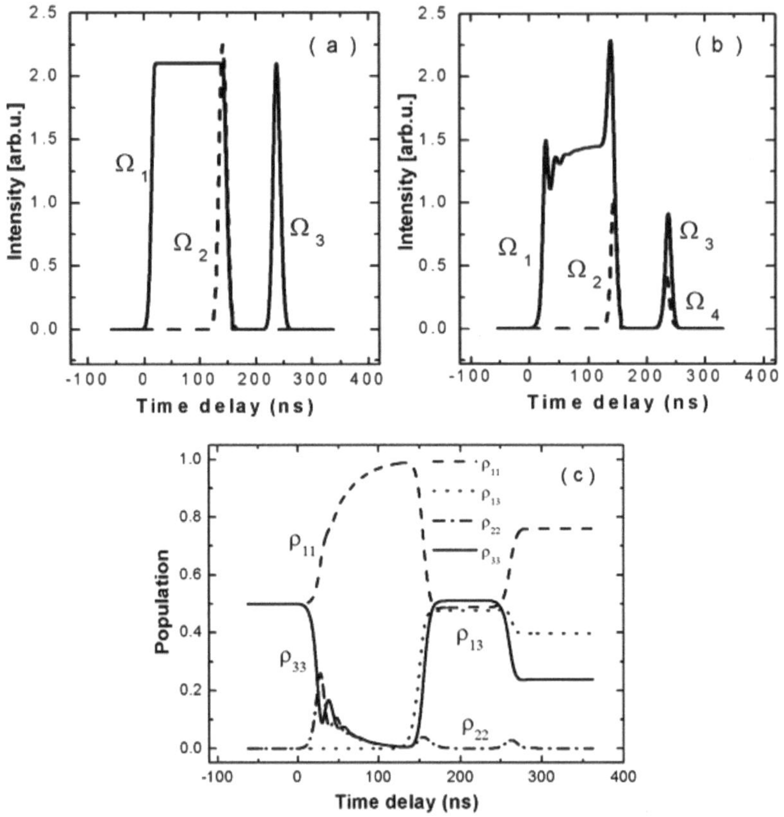

Fig. 2. Numerical simulations of obtaining the released pulse with Rabi frequency Ω_4 by the retrieve control pulse with Rabi frequency Ω_3. (a) The write control pulse, the probe pulse, and the retrieve control pulse at the entrance of the sample cell. (b) The output pulses and the restored pulse at the exit of cell. (c) Population transfer of $|1\rangle$, $|2\rangle$, and $|3\rangle$. ρ_{13} is the coherence between states $|1\rangle$ and $|3\rangle$.

corresponding to atomic density of $10^{12}\,cm^{-3}$. The focus length of L5 and L6 are 30cm. After the cell, the beams pass through another $\lambda/4$ wave plate and polarizing beam splitter. They are detected by PD1 and PD2, respectively.

In Fig. 2, we performed the numerical simulations and obtained the released pulse with Rabi frequency Ω_3. Figures 2(a) and 2(b) depict the laser pulses before and after the cell, respectively. Figure 2(a) shows the input pulse envelopes with the write control pulse preceding the probe pulse and delayed retrieve control pulse behind of them. The write control pulse and probe pulse with the same edge prepare the maximum coherence between hyperfine levels. Figure 2(b) shows the population transfer of $|1\rangle$, $|2\rangle$, and $|3\rangle$ as a function of time. ρ_{13} is the coherence term of lower states $|1\rangle$ and $|3\rangle$ .It reaches the maximum value when states $|1\rangle$ and $|3\rangle$ are of the same populations.

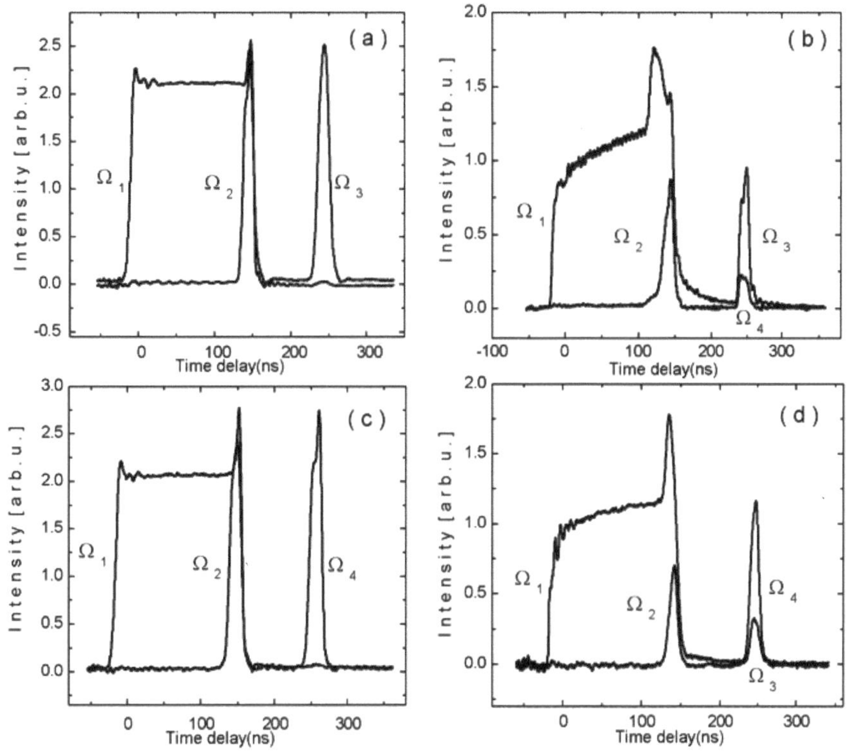

Fig. 3. (a) and (b) are the experimental results of obtaining the released pulse Ω_4 by the retrieve control pulse Ω_3. (c) and (d) are the experimental results of obtaining the released pulse Ω_3 by the retrieve control pulse Ω_4. (a) and (c) are the pulse sequences at the entrance of the Rb cell. (b) and (d) are the pulse sequences passing through the Rb cell.

The time sequences of the light storage and release processes are shown in Fig. 3. Figure 3(a) gives the experimental demonstration of the write control pulse with Rabi frequency Ω_1 $(\lambda_1 = 794.9842nm)$, the probe pulse with Rabi frequency Ω_2 $(\lambda_2 = 794.9698nm)$, and the retrieve control pulse with Rabi Frequency Ω_3 $(\lambda_1 = 794.9842nm)$ before the Rb cell. The duration of the write control pulse Ω_1 and the probe pulse Ω_2 are 170 and 30 ns, respectively. We set the edge of write control pulse being the same with the edge of probe pulse to store the latter. After storage time of 80ns, the retrieve control pulse at 794.9842nm is launched to release the stored optical pulse at 794.9698nm, as shown in Fig. 3(b). The duration of the retrieve pulse is 30ns. Similarly, we can also turn on the retrieve control field Ω_4 $(\lambda_2 = 794.9698nm)$, as shown in Fig. 3(c). Then the recovered pulse Ω_3 $(\lambda_1 = 794.9842nm)$ is released, as shown in Fig. 3(d). From Fig. 3, one can see that the write control pulse experiences absorption at the front edge of the pulse and the peak at the end of the write control pulse arises from stimulated Raman resonance scattering of the two-photon process.

B. Storage and release of light pulse based on F-STIRAP in a Pr:YSO crystal

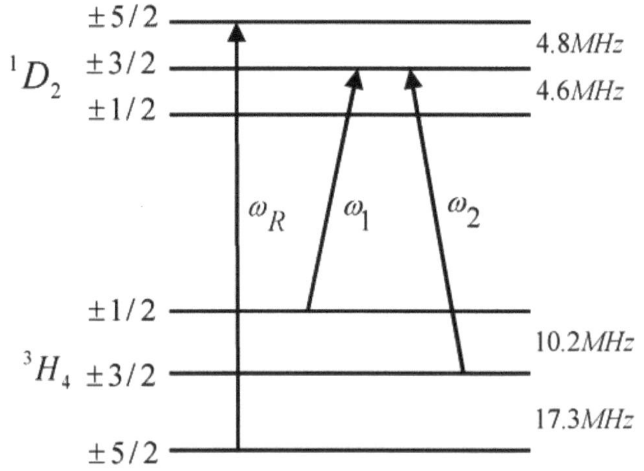

Fig. 4. The relevant energy level diagram for Pr:YSO.

The energy-level diagram of Pr:YSO is shown in Fig. 4. The crystal consists of 0.05% Pr-doped YSO in which Pr^{3+} substitutes Y^{3+}. The relevant optical transition is $^3H_4 \leftrightarrow {}^1D_2$, which has a resonant frequency of 605.977nm at site 1. The inhomogeneous width of the optical transition is about 10GHz at 1.4K, and the spin inhomogeneous width for the 10.2MHz transition is 30KHz at 1.6K [25,26]. The ground $^3H_4$ and the excited $^1D_2$ states each have three degenerate hyperfine states. We call ω_1, ω_2 and ω_R the control, signal, and repump field, respectively. The three light fields apply to a small subset of Pr ions only. The optical inhomogeneous width in this system is modified by the laser jitter due to the persistent spectral hole burning [27,35]. Then effective atomic coherence can be established at a much lower coupling intensity than would otherwise be predicted.

The waveform and the pulse sequence of the three fields are shown in Fig. 5. The process of the experiment consists of the following two steps: step-1 is the preparation of the ionic state; step-2 is the storage and release of the light pulse. The time duration of the control pulse ω_1 is $20\mu s$, and that of the signal pulse ω_2 is $5\mu s$. These two pulses have the same back edges in time. Then one more ω_1 (or ω_2) beam with $5\mu s$ duration is used as the reading pulse. The time delay between the signal pulse and the reading pulse is $15\mu s$.

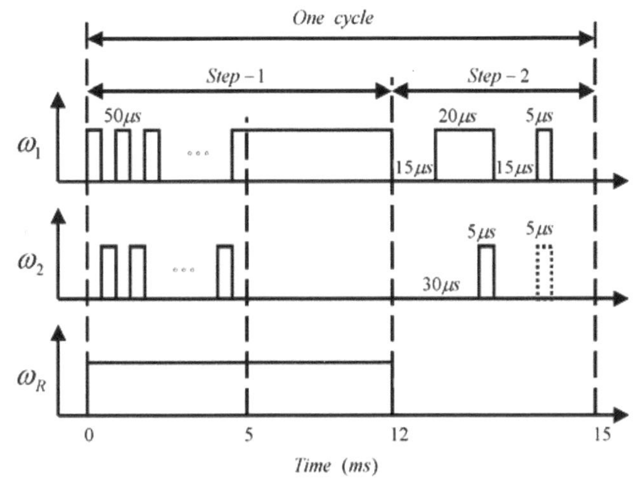

Fig. 5. Pulse sequences of the three fields in the experiment.

The experimental arrangement is illustrated in Fig. 6. A frequency stabilized dye laser (Coherent 899 ring laser) is used as the light source, and its laser jitter is about 0.5MHz. The dye laser output is split into three beams ω_1, ω_2 and ω_R by three acousto-optic modulators (AOM). The laser beams ω_1, ω_2 and ω_R are upshifted 167.9MHz, 178.1MHz and 200.2MHz from the dye laser frequency. The applied cw laser powers of ω_1, ω_2 and ω_R are 10mW, 14mW and 2mW, respectively. All three beams are linearly polarized and focused into the sample by a 30cm focal-length lens. The angle between the beams is about 10 mrad. The Pr:YSO crystal is inside a cryostat (Cryomech PT407) and the temperature is kept at 3.5K. The size of the crystal is 4mm*4mm*3mm, and optical B-axis is along 3mm. The optical signal passing through the sample is detected by a photodiode connected to a fast oscilloscope.

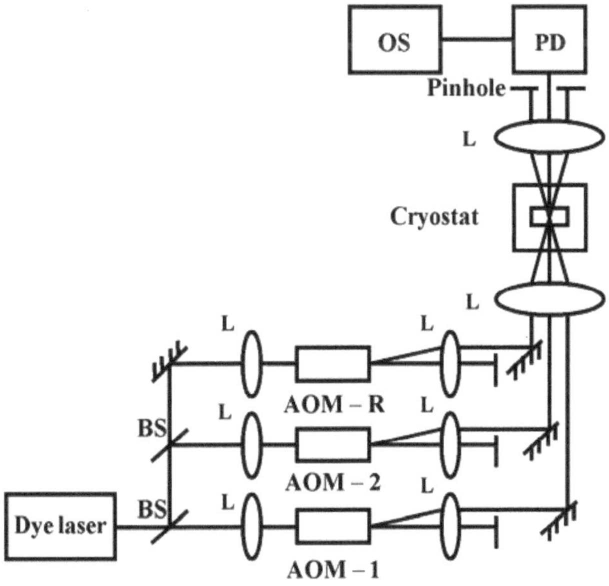

Fig. 6. Schematic diagram of the experimental setup. BS: beam splitter; L: Lens; AOM: acousto-optic

modulator; PD: photodiode; and OS: oscilloscope.

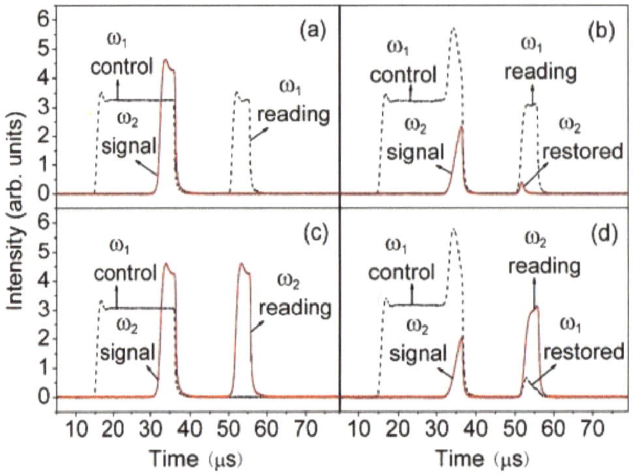

Fig. 7. Light storage and release demonstration. (a) and (c) demonstrate the pulse sequences before the crystal. (b) and (d) demonstrate the pulse sequences after the crystal.

Fig. 7 shows the experimental data of the light storage and release processes. Fig. 7(a) gives the time sequences of the three pulses before Pr:YSO crystal. We set the back edge of the control pulse being the same as that of the signal pulse to store the latter one. Due to fraction STIRAP, such two pulses excite big spin coherence between ${}^3H_4(\pm 1/2)$ and ${}^3H_4(\pm 3/2)$ levels. With the help of the control pulse ω_1, the information of the signal pulse ω_2 is stored in this coherence. After a storage time of $15\mu s$, the reading pulse ω_1 is turned on, and then the stored information of the signal pulse is released into a restored pulse with the frequency ω_2, as shown in Fig. 7(b). This restored pulse has the same frequency, polarization and propagation direction as the signal pulse. Similarly, we can turn on the reading pulse ω_2 instead of ω_1, as shown in Fig. 7(c). Then the stored information is released into a restored pulse with the frequency ω_1, as shown in Fig. 7(d). The peak at the end of the control pulse arises from stimulated emission of the two-photon process.

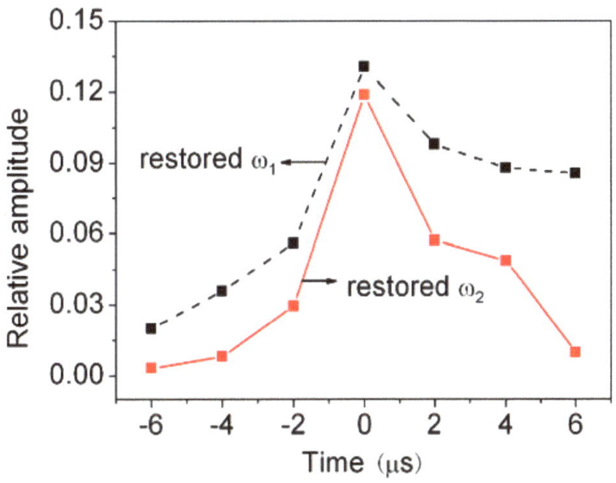

Fig. 8. The relative amplitude of the restored pulse against time delay between the control pulse and signal pulse. Zero time delay corresponds to F-STIRAP.

The peak power of the restored pulse increases with increasing peak power of the reading pulse. But the ratio of the peak power of restored pulse to that of the reading pulse, i.e. the relative amplitude of the restored pulse, is kept constant. The relative amplitude of the restored pulse is determined by the atomic spin coherence created by the control pulse and signal pulse. The bigger spin coherence corresponds to the bigger peak power of the restored pulse. In Fig. 8, we plot the relative amplitude of the restored pulse as a function of time delay between the control pulse and the signal pulse. Zero time delay corresponds the situation where these two pulses have the same back edges in time. This is the condition for obtaining the effective F-STIRAP and big atomic coherence. As seen from Fig. 8, for zero time delay, the big peak power of the restored pulse is obtained. When the time delay increases, the peak power of the restored pulse becomes small.

Applications of light storage based on EIT in a Pr:YSO crystal

Light storage based on EIT can realize the quantum state transfer between light and matter, and provides a new way to manipulate the behavior of light. It is very important in quantum information and quantum computing. Most experiments on light storage based on EIT are reported in atomic gases. Compared with atomic gases, the solid material has many obvious advantages. For practical application, the experimental researches in solid materials are more useful and important. In this section, we present the applications of light storage based on EIT in a Pr:YSO crystal.

A. Erasure of stored optical information in a Pr:YSO crystal

Long storage time is necessary for the practical applications of light storage. By employing rephasing pulses and dynamic decoherence control technique, relative long storage time has been obtained in $Pr^{3+}:Y_2SiO_5$ (Pr:YSO) crystal [35,36]. On the other hand, the erasure of stored information is also very important for some practical applications. For example, if a wrong or useless information is stored, one needs to erase the stored information quickly for the next operation. Here, we report an experimental demonstration of the erasure of stored optical information in Pr:YSO crystal. By adiabatically switching off the control field, a weak probe pulse could be stored in atomic spin coherence, and then be subsequently released by the reverse process.

By applying an erasing pulse to destroy atomic spin coherence during light storage, the stored optical information can be erased in a controlled fashion. Erasing efficiency of about 85% is observed. This erasure of stored optical information may have practical applications in information processing and all-optical network..

Figure 9 shows an energy-level diagram of Pr:YSO. The crystal consists of 0.05% Pr-doped YSO. The relevant optical transition is ${}^3H_4 \rightarrow {}^1D_2$, which has a resonant frequency of 605.977nm at site 1. We call $\omega_p, \omega_c, \omega_e$ *and* ω_r the probe, control, erasing, and repump field, respectively.

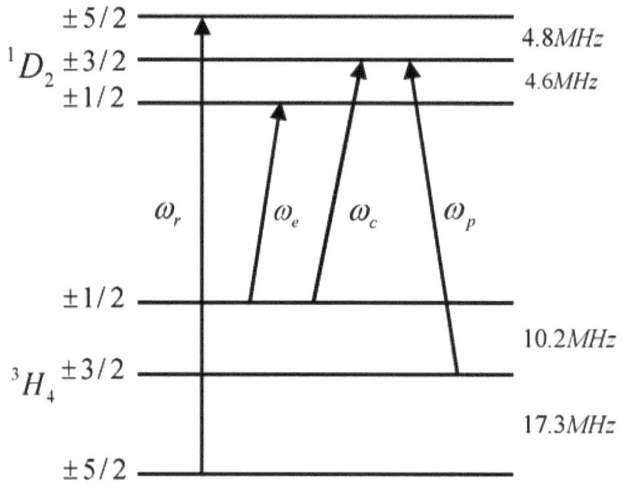

Fig. 9. The relevant energy level diagram for Pr:YSO.

The experimental arrangement is illustrated in Fig.10. A Coherent-899 ring laser (R6G dye) is used as the light source, which has about 0.5 MHz laser jitter. The laser output is split into four beams $\omega_p, \omega_c, \omega_e$ *and* ω_r by four acousto-optic modulators (AOM). The applied cw laser powers of ω_p, ω_c *and* ω_r are 3mW, 13mW and 0.5mW, respectively. The laser beams $\omega_p, \omega_c, \omega_e$ *and* ω_r are upshifted 178.1MHz, 167.9MHz, 163.1MHz and 200.2MHz from the dye laser frequency, respectively. All four beams are linearly polarized and focused into the sample by a 30-cm focal-length lens. The angle between the beams is about 10 mrad. The Pr:YSO crystal is placed inside a cryostat (Cryomech PT407) and the temperature is kept at 3.5K. The size of the crystal is 4mm*4mm*3mm, and optical B-axis is along 3 mm. The laser propagation direction is almost parallel to the optical axis.

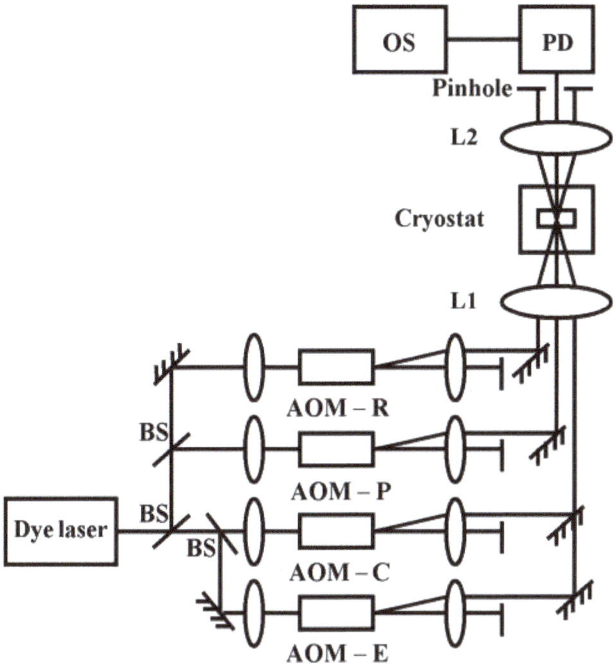

Fig. 10. Schematic diagram of the experimental setup. BS: beam splitter; L: Lens; AOM: acousto-optic modulator; PD: photodiode; and OS: oscilloscope.

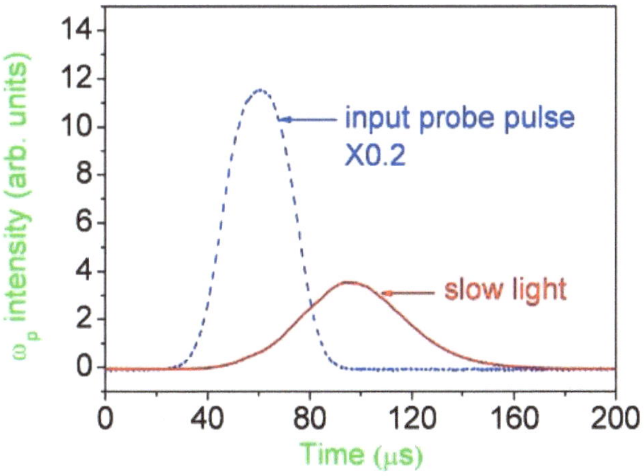

Fig. 11. Slow light demonstration

For the preparation of experimental demonstration, the ions are first populated on $^3H_4(\pm 3/2)$ level. The slowing of probe pulse is shown in Fig.11. The control field is turned on $7\mu s$ prior to the probe pulse. The dash line corresponds to the input probe pulse in the absence of the other fields. A gauss probe pulse is used to demonstrate slow light and light storage, and its 1/e full widths is $43\mu s$. Under EIT condition, the probe pulse is slowed because of the action of the control field ω_c. A time delay of about $36\mu s$ is measured from the center of the input pulse to the center of the slowed pulse.

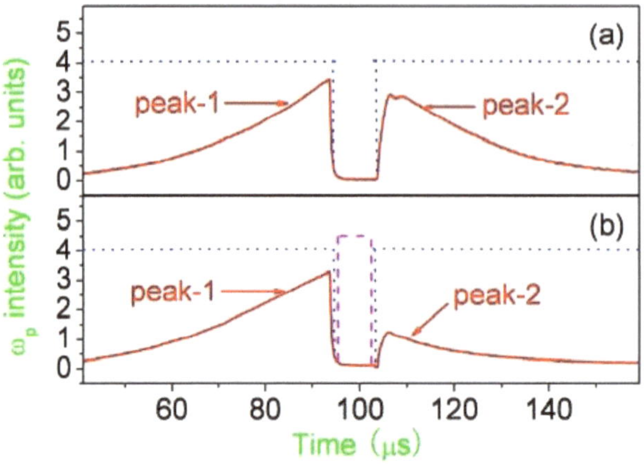

Fig. 12. (a) The probe signal in the absence of the erasing pulse. (c) The probe signal in the presence of the erasing pulse. The dot (dash) line corresponds to the control (erasing) pulse. The peak power of the erasing pulse is 25mw and its width is 9us.

When most part of the slowed probe pulse is contained in the crystal, the probe pulse is stored in the crystal by switching off the control field adiabatically. As showed in Fig.12(a), the peak-1 is the portion of the probe pulse that has left the crystal before the control field is switched off, which resulted in an observed signal that was not affected by the storage operation. The gap between peak-1 and peak-2 is the storage time of $10\mu s$. The observed peak-2 is the portion of the probe pulse that was stored in and subsequently released from the crystal by switching on the control field. In light storage process, a weak spin coherence is created when the control field is switched off. It is this coherence that the optical information of the probe pulse could be stored in. During storage time of $10\mu s$, if a square erasing pulse is applied to the crystal, which drives the transition of ${}^3H_4(\pm 1/2) \leftrightarrow {}^1D_2(\pm 1/2)$, the created spin coherence can be destroyed partially. The breakage of atomic spin coherence leads to the erasure of stored optical information. After $10\mu s$ storage time, the energy of the retrieval probe pulse is reduced remarkably due to the erasing operation, as shown in Fig.12(b). So we realize the erasure of stored optical information in a controlled fashion by applying an erasing pulse to destroy atomic spin coherence.

We define the erasing efficiency E=1-R, where R is the ratio of the retrieval probe energy with and without the erasing pulse for a fixed storage time. The attenuation of spin coherence caused by the decoherence is $\exp(-AT)$, where A is the decay constant, and T is the storage time. The attenuation of spin coherence caused by a light pulse is $\exp(-\Omega^2\tau/(2\Gamma))$, where Ω and τ are the Rabi frequency and the presence time of the light pulse, and Γ is the angular linewidth of the light transition [14,41]. For our experimental system, the attenuation of spin coherence is

$\exp(-AT)$ in the absence of the erasing pulse, and $\exp[-AT-\Omega^2\tau/(2\Gamma)]$ in the presence of the erasing pulse. For a fixed storage time, the ratio of residual spin coherence with and without the erasing pulse is $\exp(-\Omega^2\tau/(2\Gamma))$, which determines the ratio R of the retrieval probe energy with and without the erasing pulse. So $R=\exp(-\Omega^2\tau/\Gamma)$, and then $E=1-\exp(-\Omega^2\tau/\Gamma)$. From the expression of the erasing efficiency, it is seen that the erasing efficiency is determined by $\Omega^2\tau$ i. e. the energy of the erasing pulse, and does not depend on the storage time. For $10\mu s$ storage time, we measure the erasing efficiency by varying the width and intensity of the erasing pulse, as shown in Fig.13. The erasing efficiency increases with the width and intensity of the erasing pulse. We can see that, as long as the energy of the erasing pulse is constant, a shorter and stronger erasing pulse results in the same erasing efficiency as a longer and weaker erasing pulse. The experimental data are fitted with the function of $y(x)=1-\exp(-\alpha x)$, where α is the fitting parameter. The best fit is shown as solid curve in Fig.13. The experimental results are consistent with the theoretical simulation. In our experimental results, an erasing efficiency of about 85% is observed.

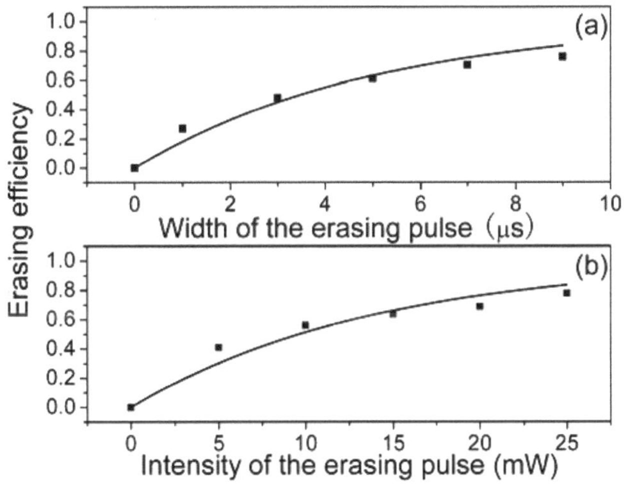

Fig. 13. (a) Erasing efficiency vs the width of the erasing pulse with a constant intensity 25mW. (b) Erasing efficiency vs the intensity of the erasing pulse with a constant width of 9us. Solid squares are experimental data. Solid curve is the theoretical fitting curve.

B. All-optical routing by light storage in a Pr:YSO crystal

In quantum information and all-optical network, it is necessary to have devices such as quantum repeater and all-optical routing, where optical information can be transferred and distributed in a controlled fashion between different light channels. Routing and wavelength-division multiplexing of optical information has many practical applications, for example, it can be used to interface optical communication line of different wavelengths and distribute optical information between different light channels [42,43]. Here, we experimentally demonstrate an all-optical routing based on light storage in a $Pr^{3+}:Y_2SiO_5$ crystal. Under the condition of EIT, the optical information of the probe pulse is stored in the crystal. By simultaneously switching on two retrieve control fields in the release process, the original optical information carried by one light channel is distributed into two light channels. This all-optical routing may have practical application in the quantum information processing and all-optical network.

Figure 14 shows an energy-level diagram of Pr:YSO. The crystal consists of 0.05% Pr-doped YSO. The relevant optical transition is ${}^3H_4 \rightarrow {}^1D_2$, which has a resonant wavelength of 605.977nm. We call $\omega_{p1}, \omega_{c1}, \omega_{c2}$ *and* ω_r the probe, control-1, control-2, and repump field, respectively.

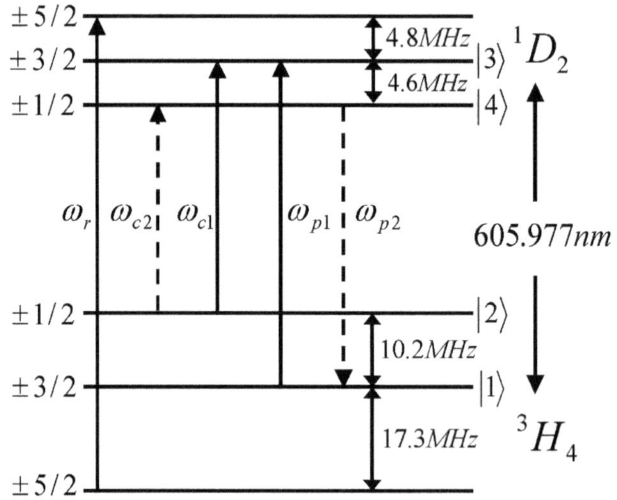

Fig. 14. The relevant energy level diagram for Pr:YSO.

The experimental arrangement is illustrated in Fig. 15. A Coherent-899 ring laser (R6G dye) is used as the light source. The laser output is split into four beams $\omega_{p1}, \omega_{c1}, \omega_{c2}$ *and* ω_r. Acousto-optic modulators (AOM) are used to upshift the frequency. The applied cw laser powers of ω_{p1}, ω_{c1} *and* ω_r are 0.7mW, 3.3mW and 0.5mW, respectively. The laser beams $\omega_{p1}, \omega_{c1}, \omega_{c2}$ *and* ω_r are upshifted 200MHz, 189.8MHz, 185.2MHz and 222.1MHz from the dye laser frequency, respectively. All four beams are linearly polarized and focused into the sample with the angle of about 10 mrad. The alignments of laser beams ω_{p1}, ω_{c1} and ω_{c2} satisfy the phase-matching condition ($\vec{K}_{p2} = \vec{K}_{p1} + \vec{K}_{c2} - \vec{K}_{c1}$) for the generation of ω_{p2} at the position indicated on L2. The Pr:YSO crystal is placed inside a cryostat (Cryomech PT407) and the temperature is kept at 3.5K. The size of the crystal is 4mm*4mm*3mm, and optical B-axis is along 3 mm.

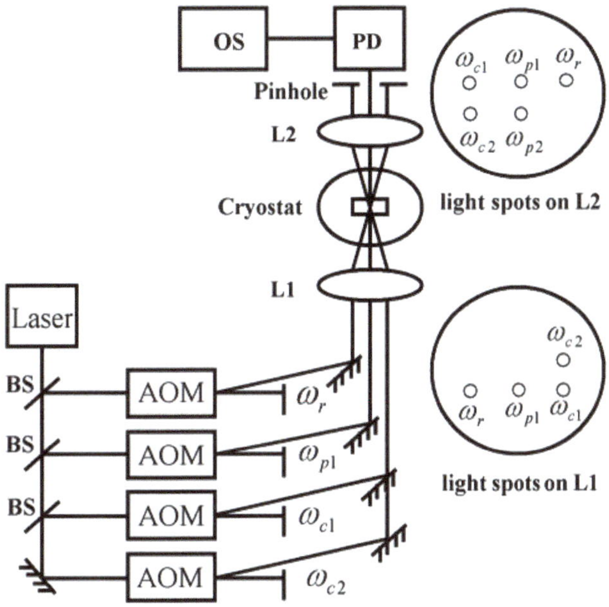

Fig. 15. Schematic diagram of the experimental setup. BS: beam splitter; L: Lens; AOM: acousto-optic modulator; PD: photodiode; and OS: oscilloscope.

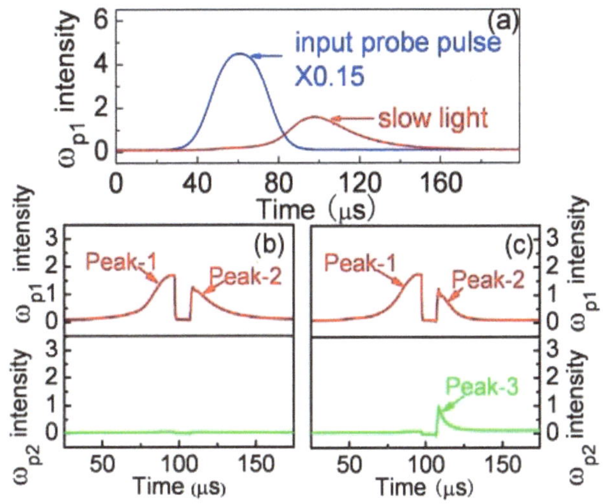

Fig. 16. (a) Slow light demonstration. (b) and (c) All-optical routing based on light storage.

For the experimental preparation of all-optical routing, we prepare the populations on $^3H_4(\pm 3/2)$ level. A gauss probe pulse is used to demonstrate light storage, and its 1/e full widths is $43\mu s$. In the step of slow light demonstration, the control-1 and repump field is applied to the crystal, and the control-2 field is not applied. The probe pulse is slowed because of the EIT effect. A time delay of about $37\mu s$ is measured, as shown in Fig. 16(a). Based on the slow light, the storage and release of the probe pulse is realized by switching off and on the control-1 field.

Figure 16(b) shows a typical light storage based on EIT. The peak-1 is the portion of the probe pulse ω_{p1} that has left the crystal before the control-1 field ω_{c1} is switched off, which is not affected by the storage operation. The peak-2 is the portion of the probe pulse ω_{p1} that is stored in and subsequently released from the crystal after switching back on the control-1 field ω_{c1}. The gap between peak-1 and peak-2 is the storage time of $10\mu s$. Because only one retrieve control field ω_{c1} is switched on in the release process, the optical information is distributed into the original light frequency ω_{p1}, and no signal of the frequency ω_{p2} is observed. Fig. 16(c) shows the all-optical routing by light storage. In order to distribute the stored optical information into two light channels, we simultaneously switch on two retrieve control fields (ω_{c1}, ω_{c2}) in the release process. In this case, we can see that the stored optical information is released at two different frequencies (ω_{p1}, ω_{p2}). Note that the released signal ω_{p2} has a different propagation direction and different carried frequency compared with the released signal ω_{p1}. The optical information originally carried by one light channel is distributed into two light channels, and then all-optical routing (or beam splitter) is realized. Especially, simultaneous switch-on of the two retrieve control fields can ensure the simultaneous transmission of the optical information at two light channels.

The shapes of two released signals ω_{p1} and ω_{p2} depend strongly on the intensity of the control-2 field. Fig. 17(a) and (b) shows the energy and temporary width of two released signals versus the intensity of the control-2 field. It is found that the energy of the released signal ω_{p2} increases with the increment of the intensity of the control-2 field, however the energy of the released signal ω_{p1} decreases with the increment of the intensity of the control-2 field. This is because that the intensity of the released signal is proportional to that of the associated control field [42]. The increment of the intensity of the control-2 field leads the result that the signal with more energy is released into the light channel with frequency ω_{p2}. So the intensity of the associated control field can control the distributing ratio of the signals between different light channels. The temporal width of two released signals decreases with the increment of the intensity of the control-2 field. The width of the released signal is inversely proportional to the spectral width of the EIT windows [44]. When the control-2 field is switched on, the original system becomes a four-level double-lambda atomic system, where the width of the EIT widows is determined by the sum of the squares of all control Rabi frequencies. So the increment of the intensity of the control-2 field leads to the decrement of the temporal width of two released signals.

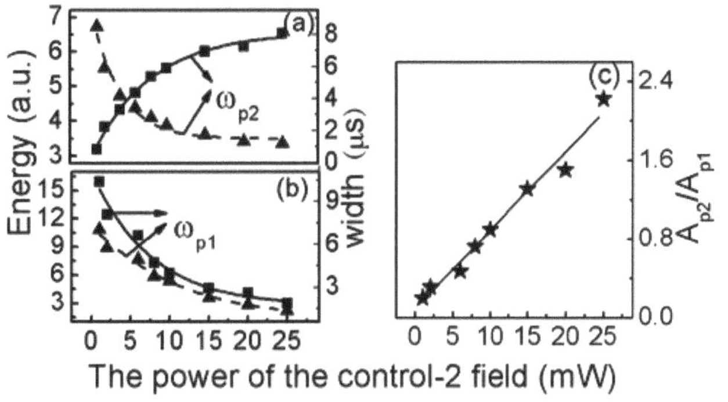

Fig. 17. (a) and (b) The energy and temporal width of the released signals vs the intensity of the control-2 field. The squares and triangles correspond to the measured energy and width of the released signals. (c) The energy ratio of the two released signals. The asterisks correspond to the measured energy ratio.

We study the energy ratio of the released signals in the two light channels, by varying the intensity of the retrieve control-2 field and keeping the intensity of the retrieve control-1 field constant. In such an EIT four-level double-lambda atomic system, the intensity of each released signal is linearly proportional to that of the associated retrieve control field. So the energy ratio (A_{p2}/A_{p1}) of the released signals is determined by the intensity ratio (I_{C2}/I_{C1}) of the corresponding retrieve control fields. Fig. 17(c) shows the energy ratio (A_{p2}/A_{p1}) of the released signals ω_{p2} and ω_{p1} as a function of the intensity of the control-2 field. We can see that the energy ratio of the released ω_{p2} and ω_{p1} is linearly proportional to the intensity of the control-2 field. The increment of the control-2 intensity leads to the increment of the energy of the released ω_{p2} and decrement of the energy of the released ω_{p1}, which is consistent with the theoretical expectation.

C. Slowing and storage of double light pulses in a Pr:YSO crystal

A conventional EIT lambda-type scheme exhibits a single dark-state polariton, and only is used to realize the coherent control of a single light pulse. A quantum bit $|\psi\rangle = \alpha|0\rangle + \beta|1\rangle$ has two basis states, so that multicomponent dark-state polaritons are required in order to realize the storage and retrieval of photonic qubits in a single atomic ensemble [15]. It is found that a tripod-type scheme exhibits double dark-state polaritons, which makes it possible to slow and store double light pulses (or photonic qubits) [15,16]. Now, matched group velocity, resonant beating of stored pulses and cross-phase modulation (XPM) have been experimentally reported in the tripod-type scheme of atomic gases [15,16,45]. Here, we experimentally demonstrate the slowing and storage of double light pulses in a tripod-type scheme of the $Pr^{3+}:Y_2SiO_5$ (Pr:YSO) crystal.

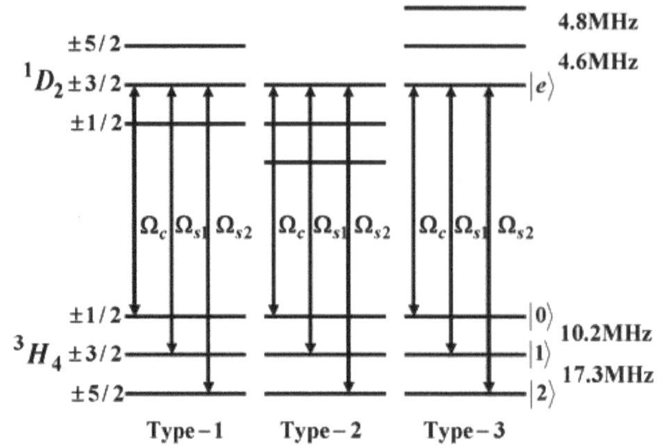

Fig. 18. The energy-level diagram of Pr:YSO.

Figure 18 shows an energy-level diagram of Pr:YSO. The crystal consists of 0.05% Pr-doped YSO. The relevant optical transition is $^3H_4 \rightarrow {}^1D_2$, which has a resonant wavelength of 605.977nm. We call Ω_{s1}, Ω_{s2} and Ω_c the signal-1, signal-2 and control field, respectively. As shown in Fig. 18, there are three types of ions having a tripod-type scheme [21], since the Pr^{3+} ions have three excited-state hyperfine levels and the optical transition is inhomogeneously broadened. $^1D_2(\pm 3/2)$, $^1D_2(\pm 5/2)$ and $^1D_2(\pm 1/2)$ is the excited state $|e\rangle$ of type-1, type-2 and type-3 ions, respectively. The tripod-type scheme exhibits double dark-state polaritons ψ_1 and ψ_2 [5,15]:

$$\psi_1(z,t) = \cos\theta(t)\Omega_1(z,t) - \sin\theta(t)\sqrt{\kappa}\rho_{01}(z,t),$$

$$\psi_2(z,t) = \cos\theta(t)\Omega_2(z,t) - \sin\theta(t)\sqrt{\kappa}\rho_{02}(z,t),$$

$$\cos\theta(t) = \frac{\Omega_c(t)}{\sqrt{\Omega_c^2(t)+\kappa}}, \quad \sin\theta(t) = \frac{\sqrt{\kappa}}{\sqrt{\Omega_c^2(t)+\kappa}}. \qquad (5)$$

Here ρ_{01} (ρ_{02}) is the atomic coherence between states $|0\rangle$ and $|1\rangle$ ($|0\rangle$ and $|2\rangle$). $\kappa = 3n\lambda^2\gamma_r c/8\pi$, where n is the ion density, λ is the wavelength and γ_r is the natural linewidth of the optical transition, and c is the free-space speed of light. The double dark-state polaritons make it possible to slow and store double light pulses in the tripod-type scheme.

The experimental arrangement is illustrated in Fig. 19. A Coherent-899 ring laser (R6G dye) is used as the light source. The laser output is split into three beams Ω_{s1}, Ω_{s2} and Ω_c. The applied cw laser powers of Ω_{s1}, Ω_{s2} and Ω_c are 1.0mW, 1.1mW and 6.0mW, respectively.

To match the levels in Fig. 1, the laser beams Ω_{s1}, Ω_{s2} and Ω_c are upshifted 200MHz, 217.3MHz and 189.8MHz from the dye laser frequency by three acousto-optic modulators (AOM), respectively. All three beams are linearly polarized and focused into the sample. The angle between the beams is about 15 mrad. The Pr:YSO crystal is placed inside a cryostat (Cryomech PT407) and the temperature is kept at 3.5K. The size of the crystal is 4mm*4mm*3mm, and optical B-axis is along 3 mm. The laser propagation direction is almost parallel to the optical axis.

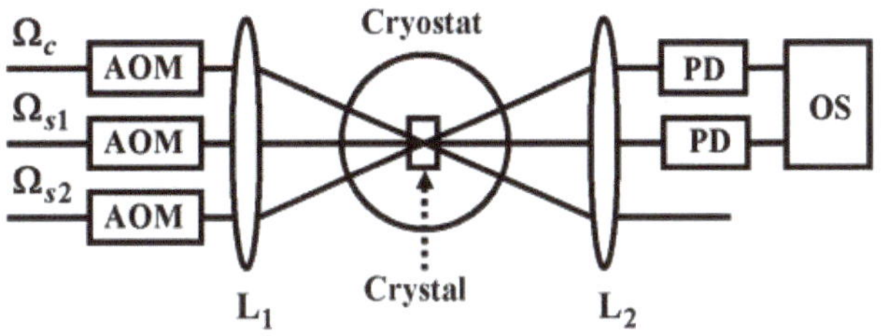

Fig. 19. Schematic diagram of the experimental setup. L, lens; AOM, acousto-optic modulator; PD, photodiode; OS, oscilloscope.

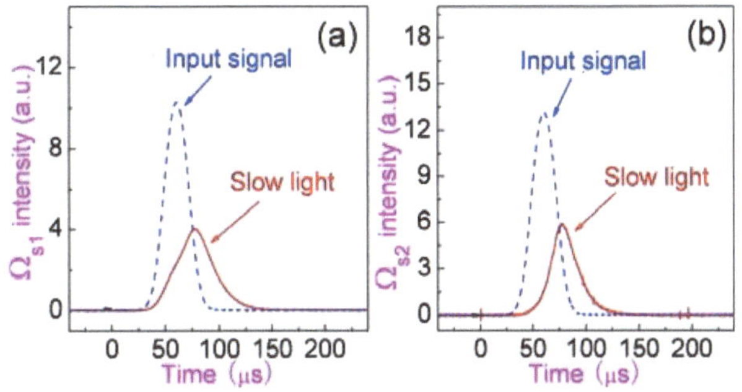

Fig. 20. Slowing of double light pulses (a) for the signal-1 and (b) for the signal-2

For states preparation, the control field is switched on for 12ms, followed by a $20\mu s$ dark time. Then the populations are pumped into levels ${}^3H_4(\pm 3/2)$ and ${}^3H_4(\pm 5/2)$ due to the optical pump. The slowing of double light pulses is investigated, as shown in Fig. 20. The dash line corresponds to the scaled-down input signal pulses. The input signal pulses are Gaussian, each with a 1/e full width of 43 microseconds, and are simultaneously generated by AOMs. In the slow light demonstration, the control field is again turned on $10\mu s$ prior to the signal pulses. Due to double dark-state polaritons of the tripod-type scheme, both signal pulses experience significant time delays. As discussed in [15], the group velocity of one signal field in tripod-type scheme can be changed by controlling the intensity of the other signal field. In our experiment, by carefully adjusting the intensities of two signal fields, these two signal pulses achieve nearly equal group velocities. A time delay of about $18\mu s$ is measured from the center of input signal pulse to the center of the slowed pulse, which corresponds to the group velocity of $V_g \approx 167 m/s$. So matched group velocities of two light pulses are obtained in a crystal with a tripod-type scheme. Double light pulses with matched group velocity have been proposed to increase nonlinear interaction time between weak pulses, and generate a large cross-phase modulation (XPM) [46], which has numerous potential applications in quantum information and communication.

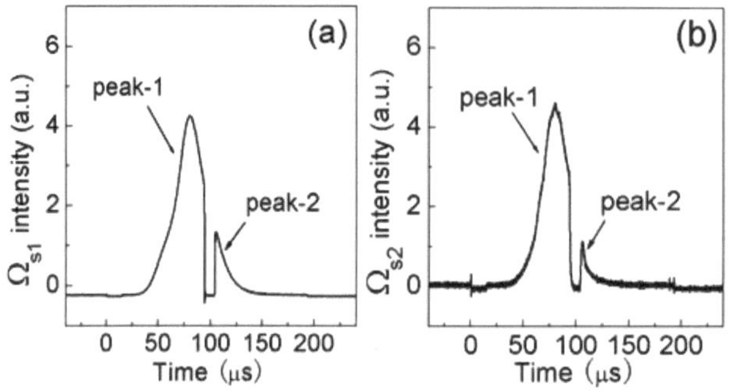

Fig. 21. Storage and retrieval of double light pulses (a) for the signal-1 and (b) for the signal-2.

According to the theory of dark-state polariton of light storage [3,15], when two slowed signal pulses are contained in the crystal, the storage and retrieval of double light pulses is realized by switching off and on the control field. When the control field is switched off, two signal pulses are converted into the ground-state spin excitation or spin coherence, which inherits the coherent optical information of these two signal pulses. When the control field is switched back on, the spin coherence is converted back into the corresponding signal field modes, and the retrieved signal pulses with the reduced group velocity continue to propagate under tripod-EIT conditions. Figure 21 shows the simultaneous storage and retrieval of double light pulses in the crystal with the EIT tripod-type scheme. The peak-1 is the portion of the signal pulses that has left the crystal before the control field is switched off, which is not affected by the storage operation. The peak-2 is the portion of the signal pulses that is stored in and subsequently released from the crystal. The gap between peak-1 and peak-2 is the storage time of $10\mu s$. The retrieved peak-2 maintains the same temporal profile as the back portion of the slow light, and its reduced intensity is caused by the dephasing of spin coherence [22]. The demonstration of the reversible storage of double light pulses paves the way towards the storage of photonic qubits, which is very useful in quantum information and quantum network.

Manipulation of atomic coherence

Atomic coherence is essential for many effects, such as electromagnetically induced transparency (EIT) [1, 2], coherence population trapping (CPT), Stimulated Raman Adiabatic Passage (STIRAP) [18-22], spontaneous emission control [47], resonant enhancement of optical nonlinearity [48,49], quantum light storage and erasure [3-5]. Most reports concentrate on the way how to generate or employ atomic coherence, such as creation of coherent superposition states [50], quantum logical gates [51], slow light storage, etc, but few discussions on the manipulation atomic coherence which has existed. In this section, we present the manipulation of atomic coherence in atomic vapor.

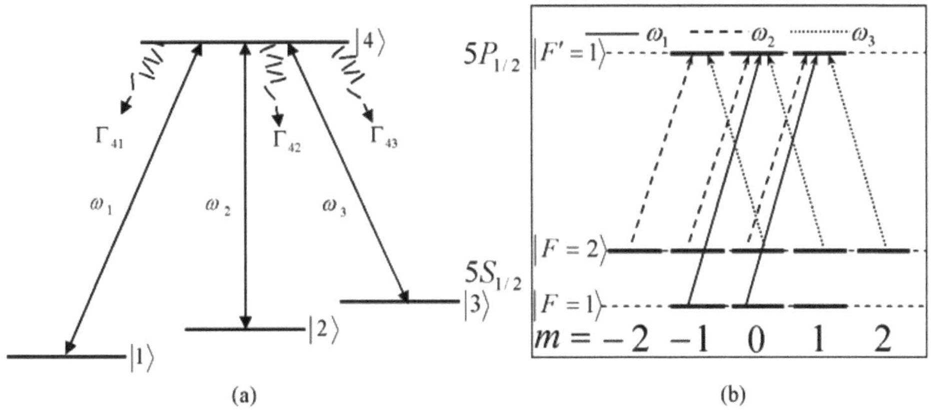

Fig. 22. (a) Theoretical model. (b) Related energy level diagram of ^{87}Rb.

In Fig. 22(a), a four-level atomic system are driven by three laser pulses with carrier frequencies of ω_1, ω_2, ω_3, and Rabi frequencies Ω_1, Ω_2, Ω_3, respectively. For simplicity, we suppose that all the populations are in the state $|1\rangle$ at the initial time, and the exited state has the same spontaneous decay rate of 5MHz to the different lower states, and ignore the dephasing rates of three lower states.

To demonstrate the time evolution process of the atomic coherence, we solve the time-dependent density-matrix equation, and the results are shown in Fig. 23. In Fig. 23, Step-1 prepares the coherence by a F-STIRAP process [52]. The system state vector could be written as: $|\alpha\rangle = \cos\theta|1\rangle - \sin\theta|2\rangle$, where the mixing angle θ is defined by the relationship: $\tan\theta = \Omega_1/\Omega_2$.

If the pulses ω_1 and ω_2 have the same time back edge, the maximal atomic coherence between $|1\rangle$ and $|2\rangle$ is reached. In step-2, we apply a STIRAP process among states $|2\rangle$, $|3\rangle$ and $|4\rangle$, the pulse sequences are shown in the right side of Fig. 23 (a1). By simulation, we find that the component of state $|2\rangle$ in the state vector $|\alpha\rangle$ is fully transferred to state $|3\rangle$ as shown in the right side of Fig. 23 (a2) and the coherence between $|1\rangle-|2\rangle$ is fully transferred to the coherence between $|1\rangle-|3\rangle$ as shown in Fig. 23 (a3). Because the states $|2\rangle$ and $|3\rangle$ are not coupled with state $|4\rangle$ in the STIRAP process, it does not arouse the stimulated emission [53] from state $|4\rangle$ to $|1\rangle$ when the pulse ω_2 is applied.

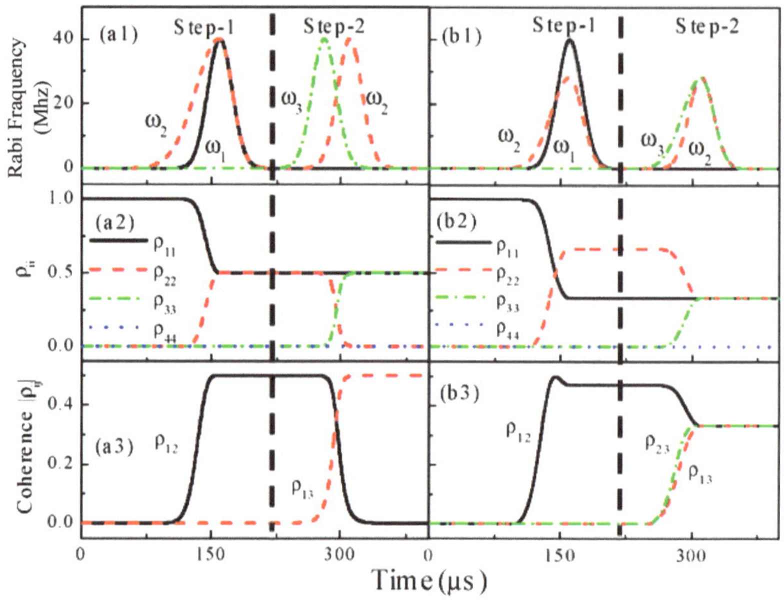

Fig. 23. The sequences of coupling pulses. (a1) the peak Rabi frequency of ω_1, ω_2 and ω_3 is 40 MHz, (b1) the peak Rabi frequency of ω_1, ω_2 and ω_3 is 40MHz, 28MHz, 28MHz, respectively, (a2b2) the square modulus of atomic states probability amplitude, (a3b3) coherence varies with time.

Next we will show how to control the coherence distribution as shown in Fig. 23 (b). By calculation, we can see that the coherence amplitude in Fig. 23 (b) step-1 is:

$$|\rho_{12}| = |\cos\theta \sin\theta| \qquad (6)$$

Here we set: tan $\phi = \Omega_2 / \Omega_3$. The coherence amplitudes in the step-2 of Fig. 23 (b) are:

$$|\rho_{12}| = |\cos\theta \sin\theta \cos\phi| \qquad (7a)$$

$$|\rho_{13}| = |\cos\theta \sin\theta \sin\phi| \qquad (7b)$$

$$|\rho_{23}| = |\sin^2\theta \sin\phi| \qquad (7c)$$

Here we set the pulse ω_2 has the same peak intensity in two steps which is easily realized in experiment.

From equations (6) and (7), we find that the atomic coherence is determined by the mixing angles θ and ϕ, which can be controlled by choosing different intensities of the input pulses. In step-1, the back edge of input pulses ω_1 and ω_2 have the same shape functions but different amplitudes, and the ratio of Ω_1 and Ω_2 is constant during the time evolution process. In step-2, the ratios of the input pulse Rabi frequencies are also constant. Ignoring the dephasing rates of the lower states, the coherence will maintain constant when the pulses vanish. From equations (6) and (7), we can regard the coherence variation process in step-2 as a coherence amplitude rotational process of step-1. The value of $\left|\rho_{12}^2+\rho_{13}^2\right|^{1/2}$ in step-2 just has the same value of $\left|\rho_{12}\right|$ in step-1. By controlling the input pulses intensity and shape, we can prepare arbitrary coherence distribution between state $|1\rangle$, $|2\rangle$ and $|1\rangle$, $|3\rangle$ which limited by $\left|\rho_{12}^2+\rho_{13}^2\right|^{1/2} \leq 0.5$ only. For example, in Fig. 23 (b1,b2,b3), the coherence is prepared between states $|1\rangle$ and $|2\rangle$, $|1\rangle$ and $|3\rangle$, $|2\rangle$ and $|3\rangle$ are the same by controlling the pulse intensity and shape in the two steps.

The above method could be expanded to N-level system, every time we choose two lower levels which couple with a common exited state by two pulses, by choosing the pulse sequences and shapes we can realize the coherence transfer and contribution in a N-level system.

We performed an experiment to verify the atomic coherence transfer process with ^{87}Rb. Fig. 22(b) shows a related level diagram of ^{87}Rb. We choose the sublevels $5S_{1/2}|F=1,m=-1\rangle$, $5S_{1/2}|F=2,m=-1\rangle$, $5S_{1/2}|F=2,m=1\rangle$, $5P_{1/2}|F'=1,m=0\rangle$ as the states $|1\rangle,|2\rangle,|3\rangle,|4\rangle$, respectively. The pulses ω_1 and ω_2 have the same circular polarization (σ^+), and the pulse ω_3 have the opposite circular polarization (σ^-). We detect the coherence by observing coherent Raman scattering signal, if there is coherence between the states $|1\rangle-|3\rangle$, when the read pulse ω_1 couple the states $|1\rangle-|4\rangle$, the coherence between $|3\rangle-|4\rangle$ will be induced. From Maxwell's equations, the coherence between $|3\rangle-|4\rangle$ will induce a stimulated signal, and the coherence along the direction of the signal propagation are the same, which ensure that the signal is enhanced continually.

The experiment process consists of the following three steps as shown in Fig. 24 (a): Step-1 prepares most atoms in the ground state $5S_{1/2}|F=1\rangle$ by an optical pumping process. Step-2 prepares the coherence. In this step, the pulse duration of ω_1 is 70ns, and that of ω_2 is 160ns. Step-3 is the coherence transfer and read process. In step-3, the pulse duration of ω_2 is 30ns which is movable, and that of the pulse ω_3 is 60ns which is fixed. When the back edge of the pulse ω_2 cross with the front edge of the pulse ω_3 and the STIRAP condition is fulfilled, the coherence between $|1\rangle-|2\rangle$ will be transferred to the coherence between $|1\rangle-|3\rangle$. In this step we call the pulses ω_2 and ω_3 the transfer pulses. After this process, one more pulse ω_1 with 40ns duration is used as the read pulse, if the atomic coherence transfer is realized, a σ^- polarized coherent Raman scattering signal will be observed. The time delay between step-2 and the reading pulse is about 200ns.

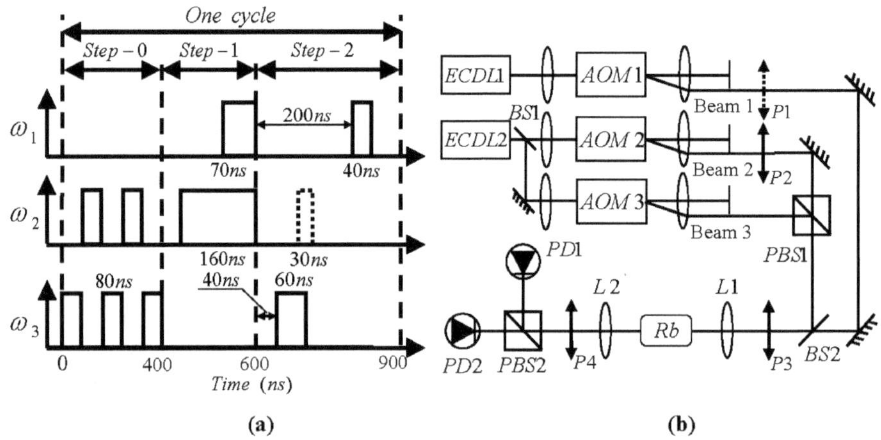

Fig. 24. (a) Pulse sequences of the three fields in the experiment. (b) Experimental setup. AOM: acousto-optic modulator, L: lens, P1,P2: $\lambda/2$ wave plate, P3,P4: $\lambda/4$ wave plate, PBS: Polarzing beam splitter, BS: beam splitter, PD: Photodiode.

The step-2 can also create the coherence between $5S_{1/2}|F=1,m=0\rangle$ and $5S_{1/2}|F=2,m=0\rangle$, and the first half of transfer pulse ω_3 can be considered as a read pulse to this coherence. As a result, the coherence is transferred to a signal with ω_1 frequency and σ^- polarizations, and the pulse ω_3 and the generated pulse will create a adiabatic passage, which causes most populations go to $5S_{1/2}|F=1,m=0\rangle$, a little populations go to other sublevels by fluorescence loss from $5P_{1/2}|F'=1,m=-1\rangle$. The populations do not affect the major conclusion of the experiment.

The experimental arrangement is illustrated in Fig. 24(b). Both ECDL1 and ECDL2 are the DL-100 external-cavity diode lasers with linearly polarized output beams. Laser from ECDL1 is tuned to the transition of ^{87}Rb $5S_{1/2}|F=1\rangle$ to $5P_{1/2}|F'=1\rangle$, and that from ECDL2 is tuned to the transition of ^{87}Rb $5S_{1/2}|F=2\rangle$ to $5P_{1/2}|F'=1\rangle$. The applied cw laser powers of this three beams passing through the cell are 6.4mW, 3.2mW and 3.2mW, respectively, and the beam diameters are about 0.1mm. The beams pass through three acousto-optic modulators (AOM) respectively driven by pulses with adjustable duration and delay (the rise time of the AOM is about 16ns). The P1 wave plates makes the beams 1 and 2 have opposite circular polarization with beam 3, and the P1 wave plate in beam 1 is movable. The focus length of L1 and L2 are 30cm. Atomic Rb vapor is contained in 3.5cm long and 2.5cm diameter glass cell, and the temperature is about 90^oC, corresponding to atomic density of $10^{12}cm^{-3}$.

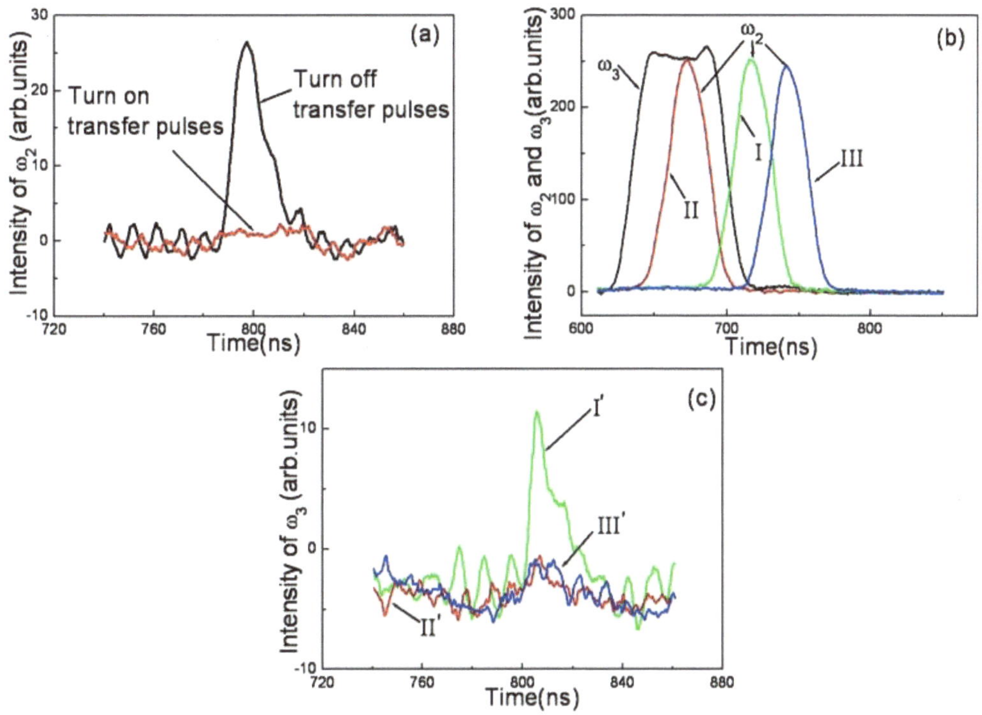

Fig. 25. (a) Generated signal fields of ω_2. (b) The transfer pulse ω_3 is fixed, and ω_2 is tuned from I to II to III. (c) Generated signals I', II', and III' respectively corresponding to I, II, and III in (a).

In the experiment, we firstly need create the coherence between $|1\rangle$ and $|2\rangle$, but the pulses ω_1 and ω_2 have the same polarization, it can not be separated by PBS. P1 is removed at beginning, which will make the two pulses have opposite polarization. We adjust the sequences of these two pulses to look for the maximal coherence point where the optimal σ^+ signal will be observed with the read pulse ω_1. The transfer pulses is turned on, the σ^+ signal will disappear as shown in Fig. 25(a). It means the coherence has been transferred or disappears. P1 is put back, a σ^- signal will be observed with the read pulse ω_1. Changing the time delay of the transfer pulse ω_2

to destroy the STIRAP condition, the signal will disappear as shown in Fig. 25(b) and 25(c). Although the phenomenon is observed through two different sublevels systems, it still can prove that the coherence have been transferred. The sublevels is symmetrical, if the σ^- polarization ω_1 induce the coherence between $5S_{1/2}|F=1, m=1\rangle$ and $5S_{1/2}|F=2, m=-1\rangle$, the σ^+ polarization ω_1 could induce the coherence between $|1\rangle - |2\rangle$. There is no mechanism to create the coherence between $|1\rangle - |3\rangle$, except that the coherence transfer from $|1\rangle - |2\rangle$ to $|1\rangle - |3\rangle$.

Conclusion

In this chapter, we reviewed our recent research works on atomic coherence and optical storage. By employing maximum atomic coherence to store light information, a new technique of light storage based on F-STIRAP is realized. The applications of light storage based on EIT in a Pr^{3+}:Y_2SiO_5 crystal are developed, including erasure of stored light information, all-optical routing based on light storage, and slowing and storage of double light pulses. We also presented the control and transfer of atomic coherence by the technique of STIRAP, which is helpful to generation of superposition states and quantum logical gates. Those works may have practical application in quantum information, all-optical network, and laser spectroscopy.

Acknowledgments

We are sincerely grateful to Dr. X. L. Song and L. Wang for their previous work. This work is supported by the National Natural Sciences Foundation of China (Grant No. 10334010, 10774059), the doctoral program foundation of institution of High Education of China, and the National Basic Research Program (Grant No. 2006CB921103).

References

1. S. E. Harris. Electromagnetically induced transparency. Physics Today 1997 Jul; 50(7):36-42.

2. M. Fleischhauer, and A. Imamoglu. Electromagnetically induced transparency: Optics in coherent medium. Reviews of Modern Physics 2005 Jul 12;77(2):633-673.

3. M. Fleischhauer and M. D. Lukin. Dark-State Polaritons in Electromagnetically Induced Transparency. Physical Review Letters 2000 May 29;84(22):5094-5097.

4. C. Liu, Z. Dutton, C. H. Behroozi, and L. V. Hau. Observation of coherent optical information storage in an atomic medium using halted light pulses. Nature (London) 2001 Jan 25. 409:490-493.

5. D. F. Phillips, A. Fleischhauer, A. Mair, R. L. Walsworth, and M. D. Lukin. Storage of light in atomic vapor. Physical Review Letters 2001 Jan 29;86(5):783-786.

6. A. S. Zibrov, A. B. Matsko, O. Kocharovskaya, Y. V. Rostovtsev, G. R. Welch, and M. O. Scully. Transporting and time reversing light via atomic coherence. Physical Review Letters 2002 Feb 26;88(10):103601(4).

7. B. Wang, S. J. Li, H. B. Wu, H. Chang, H. Wang, and M. Xiao. Controlled release of stored optical pulses in an atomic ensemble into two separate photonic channels. Physical Revew A 2005

Oct 4;72(4):043801(5).

8. K. Honda, D. Akamatsu, M. Arikawa and et al .Storage and retrieval of a squeezed vacuum. Physical Review Letters 2008 Mar 3;100(9):093601(4).

9. J. Appel, E. Figueroa, D. Korystov, M. Lobino, and A. I. Lvovsky. Quantum memory for squeezed light. Physical Review Letters 2008 Mar 5;100(9):093602(4).

10. K. S. Choi, H. Deng, J. Laurat, and H. J. Kimble. Mapping photonic entanglement into and out of a quantum memory. Nature 2008 Mar 16;452(7183:)67-71.

11. M. D. Eisaman, A. Andre, F. Massou, M. Fleischhauer, A. S. Zibrov, and M. D. Lukin. Electromagnetically induced transparency with tunable single-photon pulses. Nature 2005 Dec 8; 438:837-841.

12. M. Bajcsy, A. S. zibrov, and M. D. Lukin. Stationary pulses of light in an atomic medium. Nature 2003 Dec 11;426:638-641.

13. Y. W. Lin, W. T. Liao, T. Peters, H. C. Chou, J. S. Wang, H. W. Cho, P. C. Kuan, I. A. Yu. Stationary light pulses in cold atomic media and without bragg gratings. Physical Review Letters 2009 May 28;102(21): 213601(4).

14. Y. F. Chen, C. Y. Wang, S. H. Wang, and I. A. Yu. Low-light-level cross-phase-modulation based on stored light pulses. Physical Review Letters 2006 Feb 2;96(4):043603(4).

15. L. Karpa, F. Vewinger, and M. Weitz. Resonance beating of light stored using atomic spinor polaritons. Physical Review Letters 2008 Oct 24;101(17):170406(4).

16. A. MacRae, G. Campbell, and A. I. Lvovsky. Matched slow pulses using double electromagnetically induced transparency. Optics Letters 2008 Nov 15;33(22):2659-2661.

17. Y. F. Chen, Y. C. Liu, Z. H. Tsai, S. H. Wang, and I. A. Yu. Beat-note interferometer for direct phase measurement of photonic information. Physical Revew A 2005 Sep 14;72(3):033812(4).

18. N. V. Vitanov, T. Halfmann, B. W. Shore, K. Bergmann. Laser-induced Population Transfer by Adiabatic Passage Techniques. Annual Review of Physical Chemistry 2001 Oct;52:763-809.

19. J. Klein, F. Beil, and T. Halfmann. Robust Population Transfer by Stimulated Raman Adiabatic Passage in a Pr3+:Y2SiO5 Crystal. Physical Review Letters 2007 Sep 14;99(11):113003(4).

20. M. Oberst, J. Klein, and T. Halfmann. Enhanced four-wave mixing in mercury isotopes, prepared by stark-chirped rapid adiabatic passage . Optics Communications 2006 Aug 15; 264:463-470.

21. H. Goto and K. Ichimura. Observation of coherent population transfer in a four-level tripod system with a rare-earth-metal-ion-doped crystal. Physical Revew A 2007 Mar 15;75(3):033404 (14).

22. M. Oberst, F. Vewinger, and A. I. Lvovsky. Time-resolved probing of the ground state coherence in rubidium. Optics Letters 2007 Jun 15;32(12):1755-1757.

23. X. L. Song, L. Wang, Z. H. Kang, R. Z. Zhu, X. Lin, Y. Jiang, and J. Y. Gao. Optical signal storage and switching between two wavelengths. Applied Physics Letters 2007 Aug 14;91(7): 071106 (3).

24. H. H. Wang, L. Wang, X. G. Wei, Y. J. Li, D. M. Du, Z. H. Kang, Y. Jiang, and J. Y. Gao. Storage and selective release of optical information based on fractional stimulated Raman adiabatic passage in a solid. Applied Physics Letters 2008 Jan 30;92(4):041107 (3).

25. K. Holliday, M. Croci, E. Vauthey, and U. P. Wild. Spectral hole burning and holography in an $Y_2SiO_5:Pr^{3+}$ crystal. Physical Revew B 1993 Jun 1;47(22):14741-14752.

26. R. W. Equall, R. L. Cone, and R. M. Macfarlane. Homogeneous broadening and hyperfine

structure of optical transitions in Pr^{3+}:Y_2SiO_5. Physical Revew B 1995 Aug 1; 52(6):3963-3969.

27. B. S. Ham, P. R. Hemmer, and M. S. Shahriar. Efficient electromagnetically induced transparency in a rare-earth doped crystal. Optics Communications 1997 Dec 15; 144(4-6):227-230.

28. F. Bell, J. Klein, G. Nikoghosyan, and T. Halfmann. Electromagnetically induced transparency and retrieval of light pulses in a and a level scheme in Pr3+:Y2SiO5. J. Phys. B .2008 Apr 14;41(7):074001(9).

29. B. S. Ham, M. S. Shahriar, and P. R. Hemmer. Enhancement of four-wave mixing and line narrowing by use of quantum coherence in an optically dense solid. Optics Letters 1999 Jan 15; 24(2): 86-88.

30. H. H. Wang, D. M, Du, Y. F. Fan, A. J. Li, L. Wang, X. G. Wei, Z. H. Kang, Y. Jiang, J. H. Wu, and J. Y. Gao. Enhanced four-wave mixing by atomic coherence in a Pr3+:Y2SiO5 crystal. Applied Physics Letters 2008 Dec 9;93(23):231107 (3).

31. B. S. Ham, M. S. Shahriar, and P. R. Hemmer. Enhanced nondegenerate four-wave mixing owing to electromagnetically induced transparency in a spectral hole-burning crystal. Optics Letters 1997 Aug 1;22(15):1138-1140.

32. B. S. Ham and P. R. Hemmer. coherence switching in a four-level system: quantum switching. Physical Review Letters 2000 May 1;84(18):4080-4083.

33. B. S. Ham. Observations of delayed all-optical routing in a slow-light regime. Physical Revew A 2008 Jul 30;78(1): 011808 (4).

34. H. H. Wang, A. J. Li, D. M. Du, Y. F. Fan, L. Wang, Z. H. Kang, Y. Jiang, J. H. Wu, and J. Y. Gao. All-optical routing by light storage in a Pr3+:Y2SiO5 crystal. Applied Physics Letters 2008 Dec 5; 93(22): 221112(3).

35. A. V. Turukhin, V. S. Sudarshanam, M. S. Shahriar, J. A. Musser, B. S. Ham, and P. R. Hemmer. Observation of ultraslow and stored light pulses in a solid. Physical Review Letters 2001 Dec 20;88(2):023602 (4).

36. J. J. Longdell, E. Fraval, M. J. Sellars, and N. B. Manson. Stopped light with storage times greater than one second using electromagnetically induced transparency in a solid. Physical Review Letters 2005 Aug 2;95(6):063601 (4).

37. H. H. Wang, X. G. Wei, L. Wang, Y. J. Li, D. M. Du, J. H. Wu, Z. H. Kang, Y. Jiang, and J. Y. Gao. Optical information transfer between two light channels in a Pr3+:Y2SiO5 crystal. Optics Express 2007 Nov 20;15(24):16044-16050.

38. H. H. Wang, Z. H. Kang, Y. Jiang, Y. J. Li, D. M. Du, X. G. Wei, J. H. Wu, and J. Y. Gao. Erasure of stored optical information in a Pr3+:Y2SiO5 crystal. Applied Physics Letters 2008 Jan 3; 92(1):011105(3) .

39. H. H. Wang, A. J. Li, D. M. Du, Y. F. Fan, L. Wang, Z. H. Kang, Y. Jiang, J. H. Wu, and J. Y. Gao. All-optical routing by light storage in a Pr3+:Y2SiO5 crystal. Applied Physics Letters 2008 Dec 5; 93(22): 221112(3) .

40. H. H. Wang, Y. F. Fan, R. Wang, L. Wang, D. M. Du, Z. H. Kang, Y. Jiang, J. H. Wu, and J. Y. Gao. Slowing and storage of double light pulses in a Pr3+:Y2SiO5 crystal. Optics Letters 2009 Sep 1;34(17):2596-2598.

41. L. Wang, X. L. Song, A. J. Li and et al. Coherence transfer between atomic ground states by the technique of stimulated Raman adiabatic passage. Optics Letters 2008 Oct 15;33(20):2380-2382.

42. C. Y. Wang, Y. F. Chen, S. C. Lin, W. H. Lin, P. C. Kuan, and I. A. Yu. Low-light-level all-optical switching. Optics Letters 2006 Aug 1;31(15):2350-2352.

43. J. Apple, K. P. Marzlin, and A. I. Lvovsky. Raman adiabatic transfer of optical states in multilevel atoms. Physical Revew A.2006 Jan 9; 73(1):013804(7).

44. F. Vewinger, J. Appel, E. Figueroa, And A. I. Lvovsky. Adiabatic frequency conversion of optical information in atomic vapor. Optics Letters 2007 Oct 1; 32(19):2771-2773.

45. M. D. Lukin and A. Imamoglu. Nonlinear Optics and Quantum Entanglement of Ultraslow Single Photons. Physical Review Letters 2000 Feb 14; 84(7): 1419-1422.

46. S. J. Li, X. D. Yang, X. M. Cao, C. H. Zhang, C. D. Xie, and H. Wang. Enhanced Cross-Phase Modulation Based on a Double Electromagnetically Induced Transparency in a Four-Level Tripod Atomic System. Physical Review Letters 2008 Aug 13;101(7):073602(4).

47. Z. B. Wang, K. P. Marzlin, and B. C. Sanders. Large cross-phase modulation between slow copropagating weak pulses in 87Rb. Physical Review Letters 2006 Aug 9;97(6):063901(4).

48. S. Y. Zhu, and M. O. Scully. Spectral line elimination and spontaneous emission cancellation via quantum interference. Physical Review Letters 1996 Jan 15;76(3): 88-391.

49. H. Wang, D. Goorskey, and M. Xiao. Enhanced kerr nonlinearity via atomic coherence in a three-level atomic system. Physical Review Letters 2001 Jul 26;87(7):073601(4).

50. H. Schmidt, and A. Imamoglu. Giant kerr nonlinearities obtained by electromagnetically induced transparency. Optics Letters 1996 Dec 1;21(23):1936-1938.

51. A. Karpati, Z. Kis, and P. Adam. Engineering mixed states in a degenerate four-state system. Physical Review Letters 2004 Nov 4; 93(19):193003(4).

52. Z. Kis, and F. Renzoni. Qubit rotation by stimulated Raman adiabatic passage. Physical Revew A 2002 Feb 26;65(3):032318(4).

53. P. Marte, P. Zoller, J. L. Hall. Coherent atomic mirrors and beam splitters by adiabatic passage in multilevel systems. Physical Revew A .1991 Oct 1; 44(7):R4118-R4121.

54. V. A. Sautenkov, C. Y. Ye, Y. V. Rostovtsev, G. R. Welch, and M. O. Scully. Enhancement of field generation via maximal atomic coherence prepared by fast adiabatic passage in Rb vapor. Physical Revew A 2004 Sep 21;70(3):033406(5).

55. M. Oberst, F. Vewinger, A. I. Lvovsky. Time-resolved probing of the ground state coherence in rubidium. Optics Letters 2007 Jun 15;32(12):1755-1757.

Index

A

B

C

D

E

F

G

H

I

T

V

W

www.ingramcontent.com/pod-product-compliance
Lightning Source LLC
Chambersburg PA
CBHW041707210326
41598CB00007B/565